Vanessa Minden

Functional traits of salt marsh plants

Vanessa Minden

Functional traits of salt marsh plants

Responses of morphology- and elemental- based traits to environmental constraints, trait-trait relationships and effects on ecosystem properties

Südwestdeutscher Verlag für Hochschulschriften

Imprint
Any brand names and product names mentioned in this book are subject to trademark, brand or patent protection and are trademarks or registered trademarks of their respective holders. The use of brand names, product names, common names, trade names, product descriptions etc. even without a particular marking in this work is in no way to be construed to mean that such names may be regarded as unrestricted in respect of trademark and brand protection legislation and could thus be used by anyone.

Publisher:
Südwestdeutscher Verlag für Hochschulschriften
is a trademark of
Dodo Books Indian Ocean Ltd., member of the OmniScriptum S.R.L Publishing group
str. A.Russo 15, of. 61, Chisinau-2068, Republic of Moldova Europe
Printed at: see last page
ISBN: 978-3-8381-2435-3

Zugl. / Approved by: Oldenburg, Carl von Ossietzky Universität, Dissertation, 2006

Copyright © Vanessa Minden
Copyright © 2011 Dodo Books Indian Ocean Ltd., member of the OmniScriptum S.R.L Publishing group

Contents

Contents i
Summary iii
Zusammenfassung vi
List of most important abbreviations ix
1   Preface 1
2   The functional approach: from traits to types 4
    2.1   The term 'trait' and its functionality 4
    2.2   From functional traits to functional types 5
    2.3   Linking the environment with ecosystem properties: effect and response traits and the role of biodiversity 5
    2.4   Concepts of trait-environmental relationships 7
3   Ecological Stoichiometry 11
    3.1   General introduction 11
    3.2   Elemental composition in autotrophs in relation to tissue differentiation 12
    3.3   Variability in C:N:P concentrations 13
    3.4   Environmental constraints on organismal stoichiometry 14
    3.5   Allocation patterns 16
4   Salt marshes 18
    4.1   Abiotic conditions in salt marshes and their effects on plant performance 19
    4.2   Species composition 21
    4.3   Adaptations to salt stress 22
    4.4   Ecosystem functions in salt marshes 23
5   Study design and parameter-synopsis 25
    5.1   Study sites 25
    5.2   Sampling design 26
    5.3   Species composition 27
    5.4   Abiotic parameters 27
    5.5   Plant functional traits 30
    5.6   Ecosystem properties 32
6   Trait-environmental concepts are not explicitly convertible to salt marshes 34
    6.1   Introduction 35
    6.2   Materials and methods 37
    6.3   Results 43

6.4	Discussion	48
7	Environmental constraints on the C:N:P stoichiometry of plant organs in salt marshes	53
7.1	Introduction	54
7.2	Materials and Methods	56
7.3	Results	60
7.4	Discussion	65
8	Testing the effect-response framework: key response and effect traits determining aboveground biomass of salt marshes	71
8.1	Introduction	72
8.2	Methods	76
8.3	Results	81
8.4	Discussion	86
9	Plant trait responses to the environment and effects on ecosystem properties in salt marshes	92
9.1	Introduction	93
9.2	Materials and Methods	97
9.3	Results	102
9.4	Discussion	108
10	Synthesis	112
10.1	General remarks	112
10.2	Environmental constraints and species distribution	112
10.3	Trait expressions of salt marsh plants as response to environmental conditions	113
10.4	Possible soft traits as surrogates for hard traits	114
10.5	Trait-trait relationships	117
10.6	Plant traits effecting ecosystem functioning and the role of biodiversity	119
10.7	Improvements on methodology	121
10.8	Recommendations for further analysis and research	122
References		125
Appendix		154
Acknowledgments		160

Summary

Loss of species diversity may not only be critical from an ethical and taxonomic point of view, but it may also have profound effects on the functioning of ecosystems. Species richness *per se* may indeed not be the best predictor, but rather the functional characteristics of species from a community determine ecosystem functioning most strongly.

Trait expressions can be seen as a response of the organism to abiotic and biotic conditions. Interactions between traits give information about allocation patterns, i.e. the investment in one trait on the expense of another. Finally, traits show an effect on ecosystem properties, which can be used to evaluate the importance of plant trait expressions on the maintenance of various ecosystem properties.

In this study, the responses of plant traits on environmental constraints were evaluated and the most responsive morphology- and elemental-based plant traits were identified and tested for their applicability to trait-environmental concepts. Trait-trait relationships were investigated and ultimately, the effects of plant traits on ecosystem properties were assessed and the role of biodiversity in the maintenance of those ecosystem properties was highlighted.

The study was conducted in salt marshes along the mainland coast of Northwest Germany and on the island of Mellum. Plants inhabiting salt marshes are exposed to constantly changing environmental conditions, like periodic flooding accompanied by changes in temperature, light- and CO_2-conditions or mechanical disturbance by wind and wave energy. Trait information was collected for the whole plant (including information on belowground organs) for every species of the habitat. On every plot abiotic parameters (inundation frequency, groundwater level and salinity, utilization by land-use and soil nutrient conditions) and ecosystem properties (focusing on carbon and nutrient cycling, e.g. productivity, decomposition rates) were measured.

The results of this study clearly show that trait expressions of salt marsh species were mostly influenced by two environmental gradients, which were nutrient- and water-related. Nutrient rich soil conditions (high availability of phosphorus, potassium and carbonate, low soil sand content) were found in mainland marshes, whereas species of the island grew in comparatively nutrient poor, sandy soils. Within a marsh, a water-related gradient was detected which separated lower marsh areas with high inundation frequency and high groundwater levels and salinity from the upper marsh areas with infrequent inundation and lower groundwater levels and salinity.

A distinct set of plant traits could be detected that responded most strongly to the environmental gradients. Nutrient rich sites were inhabited by species that show high canopy height, reproductive effort (RE) and stem mass fraction (SMF), as well as low tissue C:P ratios whereas plants of nutrient poor sites exhibited high leaf and root mass fractions (LMF and RMF) and high tissue N:P ratios.

Trait expressions of salt marsh species in relation to environmental conditions followed the concepts of the functional equilibrium (Brouwer 1962b) and Tilman's allocation model (1988). The physiological-ecological-amplitude concept by Scholten et al. (1987) and the leaf economics spectrum (Wright et al. 2004) could not be confirmed from the trait perspective, the first because species of the stressful lower marsh showed higher competitive ability than those of the upper marsh, and the latter because leaf trait expressions of salt marsh species were more constrained by the water-related gradient than by nutrient availability.

By evaluating relationships between elemental-based traits (C:N, C:P and N:P ratios), isometric and anisometric relationships between 'structural' (stems and belowground) and 'metabolic' organs (leaves and diaspores) could be detected. Leaves and diaspores showed a higher degree of homeostatic regulation (self-regulation), which could be ascribed to physiological regulation which seemed decoupled from environmental conditions. Contrary, C:N and N:P ratios of stems showed higher heterostatic responses to the environment.

It could be shown that various ecosystem properties were affected by different plant traits. Whole plant C:N-ratio and SLA (specific leaf area) most strongly affected standing dead biomass, whereas standing live biomass of the community was determined by stem biomass (plant trait). Total aboveground biomass responded most strongly to belowground dry mass (plant trait), which also affected aboveground net primary productivity together with canopy height. The species richness of a salt marsh was primarily determined by plants exhibiting high belowground and low aboveground dry mass as well as low leaf and stem properties (LSP, aggregated by leaf and stem dry matter content (LDMC and SDMC), stem specific density (SSD) and SLA). The latter trait also affected decomposition rates most strongly.

It was possible to attribute trait expressions responsible for the ecosystem properties to the plant species from which they derived. It could be demonstrated that various ecosystem properties depend on trait expressions of different plant species and that multifunctionality of this ecosystem is determined by the functional diversity of the habitat.

In conclusion, the most important findings of this study were that nutrient availability and water-related environmental conditions most strongly influence trait expressions of salt marsh plants. Morphology-based traits were partly different from those of other terrestrial plants, due to the response of salt marsh species to inundation frequency, groundwater level and salinity (water-related environmental conditions). Trait-trait relationships of elemental-based traits were consistent to the findings of other studies. Different plant traits were found to affect different ecosystem properties. Therefore, multifunctionality of salt marshes can only be ensured by conserving functional diversity via species diversity. Loss of species diversity and by that loss of functional diversity might have profound effects on those ecosystem properties, which might also influence the adjacent habitats in the Wadden Sea system.

Zusammenfassung

Der Rückgang von Arten wird nicht nur aus ethischen und taxonomischen Gründen mit Besorgnis beobachtet, sondern auch aufgrund seiner weitreichenden Konsequenzen auf Ökosystemfunktionen. Für diese ist Artenreichtum *per se* nicht die beste Grundlage, sondern es sind vielmehr die funktionalen Merkmale der Arten einer Gemeinschaft, die die Basis für Ökosystemfunktionen schaffen.

Eine Merkmalsausprägung kann als Antwort einer Art auf abiotische und biotische Einflussgrößen gesehen werden. Wechselwirkungen zwischen Merkmalen liefern Informationen zu Allokationsprozessen, d.h. die Investition in ein Merkmal auf Kosten eines anderen. Schließlich wirken Merkmale auf Ökosystemprozesse ein, was wiederum eine Beurteilung von Pflanzenmerkmalen hinsichtlich ihres Einflusses auf verschiedene Ökosystemprozesse zulässt.

In der vorliegenden Arbeit wurden Merkmalsausprägungen von Pflanzen als Antwort auf Umwelteinflüsse untersucht. Zudem wurden ‚morphologische-, und ‚element-basierte' Merkmale identifiziert und ihre Anwendbarkeit auf Merkmals-Umwelt Konzepte überprüft. Wechselwirkungen zwischen Pflanzenmerkmalen und Effekte von Merkmalen auf Ökosystemfunktionen wurden analysiert, wobei die besondere Rolle von Biodiversität in der Erhaltung dieser Funktionen herausgestellt wurde.

Die Arbeit wurde auf Salzwiesen entlang der nordwestdeutschen Küste sowie auf der Insel Mellum durchgeführt. Pflanzen der Salzwiesen unterliegen ständig wechselnden Umweltweinflüssen, z.B. periodischen Überflutungen mit einhergehenden Schwankungen der Temperatur, Licht und CO_2-Verhältnisse oder mechanischer Störung durch Wind- und Wellenenergie. Daten über Merkmalsausprägungen bezogen sich auf die gesamte Pflanze (inklusive unterirdische Organe) und wurden für alle Arten der Salzwiesen erhoben. Auf jeder Untersuchungsfläche wurden Informationen über abiotische Parameter gesammelt (Überflutungshäufigkeit, Grundwasserstand und –salinität, Landnutzung und Bodennährstoffgehalt) sowie Daten über Ökosystemfunktionen erfasst (mit Hinblick auf Kohlenstoff- und Nähstofffluss z.B. Produktivität, Zersetzungsraten etc.).

Die Ergebnisse der Arbeit zeigen deutlich, dass die Merkmale von Salzwiesenpflanzen vornehmlich durch zwei Umweltgradienten beeinflusst werden, nämlich durch Nährstoffe und Wasser. Die Küsten des Festlandes weisen sich durch nährstoffreiche Böden aus (mit hohen Anteilen von Phosphor, Kalium und Karbonat und niedrige Sandanteile), wohingegen die Salzwiesen der Insel vorherrschend durch nährstoffarme, sandreiche Bedingungen gekennzeichnet sind. Innerhalb der Marschen konnte ein Wassergradient aufgezeigt werden, der untere Salzeisenbereiche mit hohen Überflutungshäufigkeiten und hohen Grundwasserständen und –salinitäten von höheren Bereichen abtrennt.

Des Weiteren wurde gezeigt, dass eine Reihe von Pflanzenmerkmalen entlang der Umweltgradienten ausgeprägt werden. Nährstoffreiche Standorte werden vor allem von Pflanzen besiedelt, die hohe Kronenhöhe (canopy height), hohe ‚reproductive effort' (RE) und ‚stem mass fraction' (SMF), sowie niedrige C:P Verhältnisse im Gewebe aufweisen. Pflanzen auf nährstoffarmen Standorten sind vor allem durch hohe ‚leaf und root mass fraction' (LMF und RMF) und hohe N:P Verhältnisse in ihren Geweben gekennzeichnet.

Hinsichtlich bestehender Merkmals-Umwelt Konzepte konnte das Konzept des ‚functional equilibrium' (Brouwer 1962b) und Tilmans ‚allocation model' (1988) bestätigt werden. Das ‚physiological-ecological-amplitude concept' (Scholten et al. 1987) und das 'leaf economics spectrum' (Wright et al. 2004) wurden hingegen widerlegt. Das erste, weil Arten der stressreichen unteren Salzwiese höhere Konkurrenzkraft ausweisen als die der oberen Bereiche, und das zweite, weil die Merkmalsausprägungen der Laubblätter in Salzwiesen stärker durch den Wassergradient als durch den Nährstoffgradient beeinflusst werden.

Bei der Analyse elementbezogener Merkmale (C:N, C:P und N:P Verhältnisse) konnten isometrische und anisometrische Verhältnisse zwischen ‚strukturellen' (Stamm und unterirdische Organe) und ‚metabolischen' Organen (Laubblätter und Diasporen) festgestellt werden. Bei Laubblättern und Diasporen konnte homeostatische Regulation (Selbstregulation), die unabhängig von Umwelteinflüssen funktioniert, nachgewiesen werden. Dahingegen zeigten C:N und C:P Verhältnisse von Stämmen heterostatische Antworten auf Umwelteinflüsse.

Diese Arbeit zeigt weiterhin, dass verschiedene Ökosystemfunktionen durch unterschiedliche Pflanzenmerkmale erklärt werden können. Durch C:N Verhältnisse (der gesamten Pflanze) und ‚specific leaf area' (SLA) konnte tote Biomasse erklärt werden, wohingegen lebende Biomasse der Gemeinschaft durch Stammmasse (Merkmal) beschrieben werden konnte.

Oberirdische Biomasse (total) wurde durch unterirdische Biomasse (Merkmal) bestimmt, durch welche auch Oberirdische Nettoprimarproduktion erklärt werden konnte. Der Artenreichtum einer Salzwiese wurde vor allem von Pflanzen mit hoher unterirdischer Biomasse und geringen ‚leaf und stem properties' (LSP, aggregiert aus ‚leaf und stem dry mater content' (LDMC und SDMC), ‚stem specific density' (SSD) und SLA) bestimmt. Durch dieses Merkmal ließen sich auch die Zersetzungsraten von Pflanzenmaterial erklären.

Die Merkmale die am stärksten die Ökosystemfunktionen erklärten, wurden den Pflanzenarten zugeordnet, durch die sie ausgeprägt wurden. Dadurch konnte gezeigt werden, dass verschiedene Ökosystemfunktionen von unterschiedlichen Arten abhängen und das ‚Multifunktionalität' in Salzwiesen von der funktionalen Diversität dieses Systems bedingt wird.

Zusammenfassend sind die wichtigsten Ergebnisse dieser Arbeit, dass die Nährstoffverfügbarkeit und das Wasserregime einer Salzwiese starke abiotische Einflussfaktoren für die Merkmalsausprägungen von Pflanzenarten in diesem Habitat sind. Bedingt durch die Anpassungen an Überflutung, Grundwasserstand und –salinität (Wasserregime), zeigten die morphologischen Merkmale teilweise Unterschiede zu denen anderer terrestrischer Arten. Wechselwirkungen zwischen elementbezogenen Merkmalen stimmten mit denen anderer Arbeiten überein. Verschiedene Ökosystemfunktionen konnten durch unterschiedliche Merkmale erklärt werden. Multifunktionalität von Salzwiesen kann nur durch die Erhaltung der Diversität gewährleistet werden. Der Verlust von Arten und der damit einhergehende Wegfall von funktioneller Diversität könnte erhebliche Konsequenzen auf die untersuchten Ökosystemfunktionen haben, welches wiederum die angrenzenden Habitate des Wattenmeeres betreffen könnte.

List of most important abbreviations

Environmental parameters	
GW.sal	Groundwater salinity
GW.mean	Mean level of groundwater
Plant traits	
ADM	Aboveground dry mass (stems and leaves)
BDM	Belowground dry mass (roots and rhizomes)
RE	Reproductive effort
LMF	Leaf:mass fraction
SMF	Stem:mass fraction
RMF	Root:mass fraction
SLA	Specific leaf area
LDMC	Leaf dry matter content
SSD	Specific stem density
SDMC	Stem dry matter content
LSP	Leaf and stem properties
Mon	Monocarpic plant
CGO	Clonal growth organ
Exc	Exclusion of salt ions
Dil	Dilution of cell sap
MAOS	Morphological adaptations to osmotic stress
Ecosystem properties	
AGB	Aboveground biomass
ANPP	Aboveground net primary productivity
Statistical abbreviations	
CA	Correspondence analysis
CCA	Canonical correlation analysis
PCA	Principal component analysis
SEM	Structural equation modeling
SMA	Standardised major axis regression
Miscellaneous	
PESS	Plant ecology strategy scheme
EP	Ecosystem property
TER	Threshold element ratio
MTE	Metabolic theory of ecology
ES	Ecological stoichiometry
MHT	Mean high tide

1 Preface

The alteration of ecosystems in terms of providing land to yield goods and services is considered the main impact of human's domination on earth and land transformation is regarded to as the driving force in the loss of biodiversity (Vitousek et al. 1997). Pimm et al. (1995) estimated that rates of extinctions are on the order of 100 to 1000 times those before humanity's dominance on Earth. However, it is not certain by how many species this planet is inhabited, which makes a calculation of extinction rates problematic. Also, some ecosystems are more affected by changes than others (e.g. oceanic islands, Steadman 1995), which results in locally higher rates of extinctions.

Nevertheless, changes in biodiversity - no matter to what extent - may lead to alterations in biological processes. Diversity provides a buffer that minimizes the susceptibility of ecosystems to changes in the environment (McNaughton 1977). However, it is not species richness *per se*, but the functional characteristics of species which have impact on ecosystem functioning (Chapin et al. 2000, Hooper et al. 2005). For example it was shown that ecosystems with a higher diversity in response types showed higher recovery (i.e. higher resilience) after disturbance (Lavorel and Richardson 1999).

Some studies suggest that changes in biodiversity can have greater effects on ecosystem properties than changes in environmental conditions (van Cleve et al. 1991, Chapin et al. 2000), e.g. a single invasive species can have profound effects on ecosystem properties (Vitousek and Walker 1989, Minden et al. 2010 a,b). However, it was stated in other studies that abiotic conditions, disturbances and functional traits have a greater effect on ecosystem properties than species richness (Loreau 1998, Enquist and Niklas 2001, Wardle et al. 2003).

Plants respond to biotic and abiotic conditions they are exposed to and, as outcomes of their responses, show a variety of traits (Simberloff 1991). On the other hand, ecosystem properties are affected by trait expressions, e.g. decomposition rates are affected by the lignin content of plant tissue (Hemminga and Buth 1991). Thus, the environment shapes plant trait expressions which are in reverse affecting ecosystem properties.

Referring to this context **this thesis aims at** (i) evincing the influence of the environment on the trait expressions of salt marsh plants, (ii) identifying morphology- and elemental-based response traits and (iii) detecting traits that effect ecosystem properties of salt marshes with reference to the role of biodiversity in the maintenance of those ecosystem properties.

Thesis-outline

This thesis is divided into three main parts, of which the first is an introductory part (chapter 2 to 5) and the second addresses and discusses specific questions (chapter 6 to 9). A synthesis is given in chapter 10 which aims at centralizing the major approaches of the two previous parts with creation of scenarios of possible continuative statistical analyses and finally, comments on methodology and recommendations for further research are given.

Chapter 2 introduces the basic concepts of the functional approach and devises the definitions on which this thesis refers to. The basic ideas of the effect/response framework are given, also introductions to the concepts of trait-environmental relationships and ecosystem functions. **Chapter 3** focuses on ecological stoichiometry, with main emphasis on the elemental composition in relation to tissue differentiation and environmental constraints. The following **chapter 4** describes salt marsh ecosystems as such. It gives an introduction to environmental parameters shaping species composition and performance and deals with different adaptations of salt marsh plants as response to the most important environmental conditions. **Chapter 5** gives an overview about the study areas and methods used to obtain the data for this thesis and presents diagrams of several parameters.

Chapter 6 tests several trait-environmental concepts for their applicability to salt marsh habitats. It identifies the strongest environmental gradients and discusses the most responsive plant traits, which are mainly morphology-based traits. It compares salt marsh specific plant traits to those of other terrestrial plants and gives reason for differences between them. The response of plant traits to environmental constraints is further deepened in **chapter 7**, which focuses on traits based on the elemental composition of different plant organs. Distinct patterns of C:N:P ratios of plant organs in relation to their function and as a response to environmental constraints are detected.

The next two chapters focus on both response and effect traits. In **chapter 8** morphology-based and elemental-based plant traits are tested for their response to environmental parameters and for their effect on a single ecosystem property. Keystone response and effect traits are identified, and the existence of indirect links from abiotic parameters to ecosystem properties mediated by plant traits is pointed out. **Chapter 9** applies the effect/response framework on various ecosystem properties. It demonstrates the dependency of ecosystem functioning on different plant traits; by assigning them to different salt marsh zones, the role of biodiversity for the functioning of the ecosystem is highlighted.

Chapter 10 synthesizes the theoretical frameworks and the findings of statistical analyses. This chapter consecutively discusses environmental conditions of salt marshes and the response they produce in the traits of salt marsh plants. It also addresses the issue of effect traits and highlights the importance of biodiversity for the maintenance of ecosystem properties. Further, scenarios are created which combine parameters which had not been merged together so far, and conjectures are given about possible statistical results. Finally, deficiencies of this thesis are addressed and need for further research is given, including an outlook of what might be expected from future work.

This thesis is a compilation of original research consisting of collection of field work data, processing and statistical analysis of this data and finally writing of manuscripts. Chapter 6, 7, 8 and 9 were written together with co-authors acknowledged in the respective chapters. Statistical analyses in chapters 6, 7 and 8 include data collected by Sandra Andratschke, Hanna Timmermann and Janina Spalke. Data-collection originally aimed at the preparation of three diploma theses which were conducted in the Landscape Ecology Group of the University of Oldenburg (Spalke 2008, Timmermann 2008, Andratschke 2009).

2 The functional approach: from traits to types

One can either recognize a community as an assemblage of taxonomic units (i.e. the composition of species occurring in a certain area) or one can concentrate on the properties that the species of the community exhibit, i.e. the traits of the plants.

The first and more 'traditional' approach focuses on the taxonomy of plants and is usually performed on the species level, whereas in the second approach the properties of the species come to a fore, which can be scaled from the individual to the community (Shipley 2010).

Contrary to the taxonomic approach, the functional type approach allows the comparison of ecologically similar, although spatially distant sites. Considering the large number of different species which occur on this planet, models based on the taxonomic approach are of limited generality, as they are not applicable elsewhere (Keddy 1992). Also, taxonomically related species might show higher plasticity as response to environmental conditions than unrelated species (Duckworth et al. 2000). For example, Knox and Palmer (1995) found that functionally similar species on different mountains in Africa growing on the same altitude were taxonomically less related (based on DNA analysis) than species of different altitudes on the same mountain which were functionally less similar.

The foundation of a functional classification of species was laid as early as 300 B.C. by the Greek philosopher Theophrastus who separated plants into trees, shrubs and herbs (Morton 1981). Since then ecologists have elaborated more sophisticated classifications, in which the scaling ranges from the organismal to the community level. The most important classification of 'early modern times' is Raunkiaer's life-form system, which assumes that species' morphology is related to climatic controls (namely the position of the perennating buds relative to the ground, Raunkiaer 1907, Raunkiaer 1934). In the 1970s, Grime introduced the C-S-R triangle, which distinguishes between competitor, stress-tolerant and ruderal strategies (Grime 1974, 1979). More recently the functional approach has gained attention as a framework for predicting ecosystem responses in the face of human-induced habitat changes such as climate change (Díaz and Cabido 1997).

2.1 The term 'trait' and its functionality

In its original meaning the term 'trait' was used as predictor of organismal performance (Darwin 1859). However, nowadays it is often used to express various types of features. Eviner (2004) for example referred to both nitrogen content of litter and soil temperature as 'traits'. This ambiguous handling has ultimately led to some degree of confusion about the term itself and the concept it refers to. This problem has been addressed by Violle et al. (2007) who elaborated a well-defined designation, which has been even more simplified by Shipley (2010). He suggests that "*a trait is any measurable property of a thing or an average property from a collection of things*" and distinguishes between traits derived from organs, individuals (organisms), populations and communities. Thus an 'organismal' trait would be any measurable property of an organism or an average property from a collection of organisms, and the same applies for population and community traits.

But when is a trait a 'functional trait'? Again, different authors give diverse ideas. McGill et al. (2006) suggest to use the term when relating to organismal performance (capacity of an organism to maintain biomass over many generations), Geber and Griffen (2003) define functional trait as a "*phenotypic character that influences organismal fitness*" and Weiher et al. (1999) link it with life-history processes, i.e. germination, growth and reproduction. In this thesis 'functional traits' are defined as properties of an organism or a part of an organism which impact(s) fitness via their effects on growth, reproduction and survival (Violle et al. 2007, Shipley 2010).

2.2 From functional traits to functional types

As for the term trait, an unambiguous definition of the term 'type' has been subject to debate for some time. Root (1967) introduced the term 'guild' as a group of species that exploit the same class of environmental resources in a similar way. However, this term seemed more applicable to animal ecology, as animals actually *use* resources, whereas plants are thought of to *respond* to environmental constraints (Simberloff 1991). Other terms were introduced as well, like 'structure-functional group', 'adaptive syndrome' or 'plant type' (Wilson 1999).

It now seems that authors have agreed on 'functional types' as appropriate term, which is defined as nonphylogenetic groupings of species that share similar traits as response to environmental and biotic conditions (Wilson 1999, Duckworth et al. 2000). Differences between members within one functional type are usually much finer than those between functional types (Smith and Huston 1989).

2.3 Linking the environment with ecosystem properties: effect and response traits and the role of biodiversity

Lavorel and Garnier (2002) introduced the response-effect framework, which classifies species according to their response to the environment and their effect on ecosystem properties, respectively. Likewise, functional response groups are groups of species with a similar response to the environment, and functional effect groups those that share equivalent effects on one or more ecosystem properties (Lavorel et al. 1997, Díaz and Cabido 2001). However, some traits might show both responses and effects, and are thus keystone traits which have high contribution to the functioning and stability of ecosystems and biogeochemical cycling (Loreau et al. 2001, Lavorel and Garnier 2002). Traits may show direct responses and effects, but they may also show indirect responses and effects mediated through other traits (see Fig 2-1)

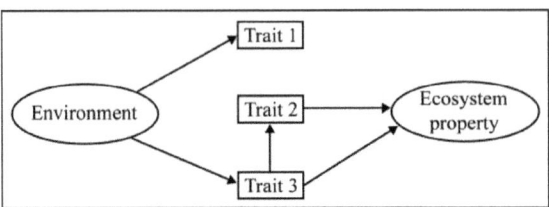

Figure 2-1: Graphical description of the effect/response framework (adapted from Lavorel and Garnier 2002). Both trait 1 and trait 3 show a direct response to the environment, whereas trait 2 is indirectly responsive (mediated through trait 3). Trait 1 does not show an effect on the ecosystem property (EP); traits 2 and 3 both show a direct effect on the EP, and the total effect of trait 3 is partly mediated through trait 2.

According to Christensen et al. (1996) ecosystem functioning includes inputs, outputs, material and energy cycling and the interactions of organisms and comprises ecosystem properties, goods and services. 'Ecosystem properties' refer to biogeochemical cycling and storage of materials like decomposition and primary productivity without any reference to a support of human life (focus of the present thesis, Hooper et al. 2005). Opposite to that, 'ecosystem goods' have direct market value (food, fresh water), whereas 'ecosystem services' have values to human welfare, but are rarely sold or bought (e.g. recreation areas, Christensen et al. 1996, Chapin et al. 2000, see Table 2-1). Estimated values of ecosystem properties can entail usage and thus cause a transition into ecosystem goods and/or services, for example when hitherto unemployed grassland provides the basis for future cattle breeding. Also, an ecosystem property like plant and animal diversity might serve as a premise for a recreation area and by that could also be considered an ecosystem service.

Table 2-1: Definitions of ecosystem properties, goods and services (see Richardson 1994, Christensen et al. 1996, Chapin et al. 2000, Hooper et al. 2005).

Ecosystem function	Definition	Example
Ecosystem property	Pools and fluxes of energy and materials within an ecosystem	Primary productivity, decomposition
Ecosystem good	Ecosystem properties with direct market value	Construction material, medicinal plants, tourism
Ecosystem service	Ecosystem properties with benefits for human endeavors	Climate regulation, aesthetic and cultural values

Traits explaining one ecosystem property might not be necessarily relevant for another (Hooper et al. 2005). One major related subject is the role of biodiversity for providing different ecosystem properties, especially in the context of its loss induced by habitat transformation. There is no doubt that global extinction of species is of major concern of conservation biology. However, local extinctions on a smaller scale or even large changes in abundance can also have profound effects on ecosystem properties (Zimov et al. 1995), for which biodiversity provides an insurance or buffer against declines (Chapin et al. 2000). Communities with higher diversity have higher probability of phenotypic trait diversity, which provides greater guarantees that functioning is consistently procured even if some species cease to exist (insurance hypothesis, Yachi and Loreau 1999, Loreau et al. 2001). Thereby redundancy of functional effect traits and diversity of functional response traits ensures ecosystem functioning (Walker et al. 1999). As long as response and effect traits are not too closely linked, loss of all species capable of performing ecosystem properties is reduced (Chapin et al. 1996, Hooper et al. 2005).

The functional type approach can be applied to various spatial scales, from a plant community, through the ecosystem and landscape to global scales (Wright et al. 2004, Kerkhoff et al. 2006, Vile et al. 2006). Actually, in order to address certain questions it is necessary to scale up from the individual to the community level. Especially ecosystem properties depend on effect traits weighted by the abundance of the species in the community (Violle et al. 2007). The 'mass ratio hypothesis' by Grime (1998) states that ecosystem properties are strongly controlled by the characteristics of the dominant species, and are relatively insensitive to the attributes of subordinate plants. However, it was demonstrated that some less abundant species also have strong effects on ecosystem properties (so-called keystone species, Paine 1969). These species might be lower in abundance, but nevertheless show large effects (Power et al. 1996); which supports the conservation of all existing species, regardless of their abundance.

2.4 Concepts of trait-environmental relationships

Plant trait expressions are a response to biotic and abiotic influences. Plant ecology strategy schemes (PESS, Westoby 1998) aim at deviating plant strategies from trait expressions in order to centralize fundamental concepts of plant ecology. However, not all PESS include environmental forces, see the leaf economics spectrum (Wright et al. 2004). One aim of this thesis was to test several PESS for their applicability on salt marshes, including both the influence of the environment and biotic interactions that shape the trait expressions of marsh plants (PESS summarized in Fig. 2-2).

2.4.1 The leaf and whole-plant economics spectrums

The leaf economics spectrum describes a strong trade-off between leaf traits related to high metabolic rates (photosynthetic assimilation rate and respiration rate, high nutrient concentration) and leaf lifespan, which includes a high correlation to SLA (specific leaf area, Wright et al. 2004). These trades describe a gradient from potentially fast growing species with high acquisition and turnover of resources to retention of biomass and nutrients of potentially stress tolerant slow growing species.

Freschet et al. (2010) expanded this concept to the so-called 'plant economics spectrum' in which they included traits of leaves, stems and roots (pH, nitrogen content, C:N ratio, dry matter content and lignin) and related them to soil nitrogen content. Their central message was that the findings of the leaf economics are applicable to other plant organs and are also consistent in a whole plant economics spectrum.

2.4.2 Physiological-ecological-amplitude-concept

This concept is most often attributed to Scholten et al. (1987), however the underlying idea has been formulated earlier by Etherington (1975) and Ernst (1983). The concept describes the distribution of organisms along strong environmental gradients, in which the boundary of a population towards the extreme end of a gradient is constrained by abiotic conditions, whereas the more 'benign' parts of the gradient are colonized by species with high competitive ability. Applied to salt marshes this means that the lower parts of the marsh are colonized by species able to tolerate stress induced by seawater (frequent inundation and waterlogging, high groundwater levels and salinities etc.), whereas the upper marsh is inhabited by species with high competitive ability. This concept has been tested and confirmed by various salt marsh specific studies (Russell et al. 1985, Bertness and Ellison 1987, Bertness 1991, Pennings and Callaway 1992, Bockelmann and Neuhaus 1999).

2.4.3 Brouwer's theory of functional equilibrium

Brenchley (1916) and Maximov (1929) found an increased biomass allocation to roots as a response of reduced belowground resources (nutrients and water) and Shirley (1929) found a similar response of shoots at decreased aboveground resources (light intensity and quality). However, when nutrients are resupplied after being withheld for some time shoot:root ratios return to the values of those of plants that received the resources all the time (Brouwer 1962a).

These observations were brought together and formulated by Brouwer (1962a) as the theory of functional equilibrium. The core of this theory is that plants shift their biomass allocation to the organs which experience a decrease of a particular resource, i.e. to shoots when aboveground resources are impaired (light and CO_2) and to roots when water and nutrients (belowground resources) are limiting. These shifts enable the plant to capture more of those resources that most strongly limit the plant's growth. Brouwer also proposed a mechanism by which the plant regulates allocation. Whereas this mechanism has been questioned (Cheeseman 1993, van der Werf and Nagel 1996), the basic predictions were not challenged.

2.4.4 Tilman's ALLOCATION model

This model states that plants that allocate their biomass to roots are best competitors when belowground resources are abated (Tilman 1988). One the other hand, best competitors for light are those plants that allocate large fractions of their biomass to aboveground organs (i.e. stems and leaves). Under 'ideal' conditions both strategies would lead to slow growth, because in the first case plants allocated their biomass away from energy acquisition towards heterotrophic roots and in the second case towards heterotrophic stems. However, when competition is low, species that allocate their biomass to leaves are favored, which would consequently lead to high relative growth rate (RGR).

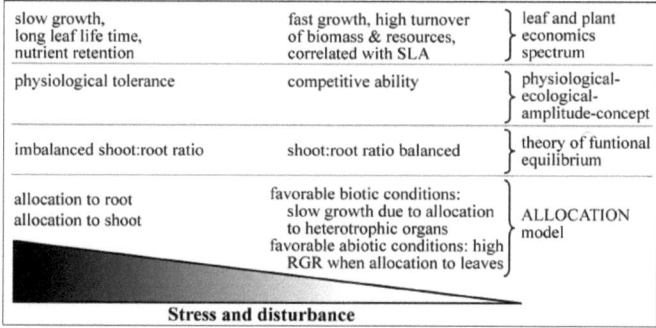

Figure 2-2: Summary of different PESS along a stress gradient.

Two concepts were not incorporated in this thesis, although considered fundamental contributions to PESS; these are **Grime's CSR scheme** and **Westoby's LHS scheme**.

Grime's CSR scheme is based on three determinants of herbaceous vegetation, that are competition (tendency of neighboring plants to utilize the same amount of light, nutrients, water and space), stress (shortages of light, nutrients, suboptimal water or space) and disturbance (impact of animals and man (grazing, trampling, ploughing), physical phenomena (soil erosion), Grime 1973, Grime 1974). Although Grime's scheme has been recognized important for the establishment of functional classifications, various aspects have also been criticized. Grime chose a triangle to illustrate his scheme and studded each corner with one of the mentioned determinants. According to his scheme a fourth corner could not be colonized, which has been vitiated later (Campbell and

Grime 1992, Burke and Grime 1996). Grubb (1985) argued that not all species show the same adaptations to a particular stress, whereas the CSR scheme is based on a consistent syndrome seen in stress-resistant species. Applied to salt marshes, some species show adaptation to salt stress by active excretion of excessive ions whereas others grow succulent in order to dilute their cell sap. Also, the scheme does not distinguish continuing from episodic disturbance (Grubb 1985) and Westoby (1998) argues that the axes are defined by reference to the concept, but as a protocol for positioning species beyond the concept is missing, it lacks transferability for worldwide comparison.

Instead, Westoby (1998) proposed a strategy scheme based on three traits (specific leaf area, plant height and seed mass; **leaf-height-seed scheme (LHS)**). The assumption is that each of these traits represents a trade-off that controls plant strategies and that it could be applicable worldwide. However, Jardim and Batalha (2008) were unable to apply the scheme when comparing strategies of animal dispersed species and those dispersed by abiotic means in a study in Brazil. Also, Golodets et al. (2009) could not prove independence between the LHS traits in a Mediterranean grazing system as proposed by Westoby, and finally Laughlin et al. (2010) criticized the neglecting of belowground resource capture strategies, which is also critical in the view of this thesis.

3 Ecological Stoichiometry

3.1 General introduction

Ecological stoichiometry studies the balance of elements and energy in ecological interactions, whereby it focuses on understanding how the balance of chemical elements and energy influences and is affected by the structure and functioning of biological systems (Elser et al. 2000b, Sterner and Elser 2002, González et al. 2010).

Organisms themselves are outcomes of chemical reactions and thus their growth and reproduction are constrained by the supply of key elements (Elser and Hamilton 2007). Plants require about 30 different elements, from which some are required in large (macroelements, e.g. H, C, N, O, Mg, P, S, K, Ca), others only in small quantities (like Mn, Zn, Fe, Mo, Ågren 2008). Especially carbon, nitrogen and phosphorus are recognized as key elements. Nitrogen and phosphorus most commonly limit plant growth and carbon provides the structural basis by constituting about 50% of a plant's dry mass and can thus also be limiting (Elser et al. 2007). The concentration of these elements in the plant's biomass is determined by the balance of uptake (nitrogen and phosphorus), assimilation of carbon, and losses due to turnover, exudation, leaching, herbivory and parasitic impacts (Chapin and Shaver 1989, Aerts and Chapin 2000). Carbon, nitrogen and phosphorus are used to synthesize various biochemical compounds crucial for growth, maintenance and reproduction (Table 3-1).

Homeostatic regulation describes the extent to which the internal elemental content is regulated in relation to the elemental supply available (Sterner and Elser 2002). Heterotrophs are considered more homeostatic than autotrophs, however also autotrophs need to regulate their elemental uptake (Güsewell 2004, Frost et al. 2005). For example excessive uptake of nitrogen and phosphorus leads to toxic effects (de Graaf et al. 1998, Lucassen et al. 2002), whereas a reduction of those two elements below the minimum to maintain cell function results in the senescence of tissue (Batten and Wardlaw 1987).

Table 3-1: Examples of chemical compounds synthesized by plants that contain carbon (C), nitrogen (N) and phosphorus (P), adapted from Sterner and Elser (2002).

Biological Material	Examples	Function / Comment
Carbohydrates: C	Glucose, starch, cellulose, lignin	Storage of energy, structure / Generally without N and P
Energetic nucleotides: C, N, P	ATP, NADPH	Energy carrier
Lipids: C, P	Phospholipids	Cell membranes
Pigments: C, N	Chlorophyll	Light adsorption / contains 6.5% N
Proteins: C, P	RUBSICO	Metabolism
Nucleic acids: C, N, P	DNA, RNA	Storage, transmission and expression of genetic information
Inorganic materials: C, P	Polyphosphate	Nutrient and energy storage

The uptake rates of particular elements are regulated by the internal store of those elements, e.g. nitrogen-deficient plants increase the rate of nitrogen uptake and decrease that of phosphorus, whereas phosphorus-deficient plants do the opposite (Aerts and Chapin 2000). Phosphorus uptake is down-regulated in response to high phosphate concentrations in the phloem (Schachtman et al. 1998). Alternatively to reduced uptake, phosphorus is stored as polyphosphate in stems, roots and seeds, which reduces the risk of toxic effects and supports growth at other times (Handreck 1997). Autotrophs are able to store relatively large quantities of nutrient elements, such as nitrogen (e.g. nitrate, amino acids) and phosphorus (e.g. phytic acid, polyphosphate or inorganic phosphate) as well as nutrient-free organic compounds (e.g. sugars) in their vacuoles. An increase in nutrients beyond what is immediately needed for growth is referred to as 'luxury consumption', which occurs when a different limiting factor (e.g. light) begins to limit growth (Sterner and Elser 2002).

3.2 Elemental composition in autotrophs in relation to tissue differentiation

The tissue differentiation of higher plants into roots, stems, leaves and later at ontogenesis reproductive organs is accompanied by a specialization of function (nutrient and water provision by roots, structural support by stems, synthesis of photo-assimilates by leaves and production of e.g. seeds or offshoot by reproductive organs). As a result, elemental composition varies between plant organs of multicellular organisms and changes through ontogenesis (Ågren 2008), but is also dependent on environmental factors (availability of nutrients and water, physiological stressors, see chapter 3.4).

Roots can take up nutrients either by interception, mass flow or diffusion. As roots elongate and grow through the soil, they sometimes intercept directly the nutrients in the soil solution (Foth and Ellis 1996). Transpiration is the major cause for mass flow: the soluble fraction of nutrients in the soil solution flows to the roots as water is taken up. By creation of a concentration gradient through the roots, plants are able to take up nutrients via diffusion, especially phosphorus and potassium (Foth and Ellis 1996). The most active tissue belowground are fine roots, which have varying C:N:P stoichiometry with generally higher nutrient content in the smallest diameter roots (Gordon and Jackson 2000).

Immature leaves grow and assimilate simultaneously, so that the demand for nutrients is defined by the stoichiometry of the biochemical processes (photosynthesis, DNA duplication and transcription, Güsewell 2004). With increasing age C:N and C:P ratios increase, as do N:P ratios, because RNA demand gets less as maturity advances (Usuda 1995, Santa Regina et al. 1997). During senescence nitrogen and phosphorus concentrations are reduced further and nutrients are usually resorbed prior to leaf shedding (Batten and Wardlaw 1987).

The elemental composition of chloroplasts shows that these organelles are rich in nitrogen, but poor in phosphorus (Kirk and Tilney-Bassett 1978). Carbon-fixation through photosynthesis is tightly coupled to nitrogen content (via Rubsico), which presupposes rRNA (phosphorus-based, Ågren 2004, Frost et al. 2005). Generally, leaves contain more nitrogen per unit dry mass than either stems or roots (Epstein 1972, Kramer and Kozlowski 1979).

Plant structural material like stems have low nitrogen and phosphorus content. They predominantly consist of lignin and cellulose, which are nutrient poor structures that contribute towards high C:N and C:P ratios in the plant's biomass (Sterner and Elser 2002).

A considerable fraction of a plant's nitrogen and phosphorus is invested in reproduction (seeds), which typically exceeds that of carbon (Güsewell 2004). Seeds show higher nitrogen and phosphorus concentrations and lower N:P ratios than vegetative structures, indicating that phosphorus limitation might affect reproduction more than nitrogen limitation (Fenner 1986). Similarly, flowering plants (or shoots) have lower N:P ratios than non-flowering plants of the same species (Eckstein and Karlsson 1997).

3.3 Variability in C:N:P concentrations

The C:nutrient concentrations of autotrophs are more variable than those of heterotrophs. One reason for this is that autotrophs acquire elements independently from each other, while heterotrophs obtain nutrients bound together via their food (Frost et al. 2005). Güsewell and Koerselman (2002) state that intraspecific variation is more important than interspecific variation when comparing nitrogen and phosphorus concentrations of wetland plants.

The natural N:P ratio of terrestrial plants at their natural site is 12-13 (Elser et al. 2000a), however individual measurements can range from 1-100 (Phoenix et al. 2003, Tomassen et al. 2004). With increasing age and size, stoichiometry at the level of the whole plant will change due to changes in the partitioning between plant tissues. Also, C:nutrient concentrations decrease during ontogenesis, because the supportive, carbon-rich and nutrient-poor tissues increase and nutrients become more 'diluted' (Kerkhoff et al. 2006, Ågren 2008). Differences in exposure to sunlight may also affect stoichiometry. Valladares et al. (2000) found that phosphorus concentrations in leaves exposed to the sun were about twice as high as those in the shade, whereas there were only small differences in nitrogen and potassium. However, Ågren (2008) assigns changing supply of nutrients through the environment and luxury consumption as major causes for stoichiometric variability.

3.4 Environmental constraints on organismal stoichiometry

3.4.1 Nutrient limitation

It is generally assumed that primary productivity is limited by nitrogen in terrestrial systems and by phosphorus in freshwater ecosystems (Schlesinger 1997), however recent work suggests equivalent nitrogen and phosphorus limitation in both systems (Elser et al. 2007). Tropical ecosystems undisturbed by glaciations are more frequently limited by phosphorus because of greater soil age, whereas temperate ecosystems may be more often limited by nitrogen (Reich and Oleksyn 2004). Low nutrient conditions lead to decreased growth and increased biomass C:nutrient ratios and more effective usage of nutrients by the plant (Vitousek 1982). In contrast, in high nutrient conditions protein synthesis and growth are maximized as competitive strategy and biomass C:nutrient ratios are decreased (Ågren 2008, Matzek and Vitousek 2009). Because of their greater photosynthetic nitrogen-use efficiency, C_4 plants can grow better at lower nitrogen supply than C_3 plants, and are therefore more advantaged on nitrogen-poor dry soils (Güsewell 2004).

To achieve a relative growth rate, a plant has to take up a minimum amount of all essential nutrients required to maintain this rate. However, availability of all these elements is often restricted. Therefore, relative growth rate is limited by the availability of the nutrient least available relative to the plant's requirements (Liebig's law of the minimum, Liebig 1840, 1855). Plants are regarded as nutrient deficient when their production (e.g. relative growth rate (RGR), tree diameter increment, annual herbal production etc.) is slower by a certain percentage (e.g. 10% or 20%) than the maximum which could be reached at high nutrient supply (Wells et al. 1986, Föhse et al. 1988). The threshold element ratio (TER) is the critical ratio at which a nutrient is limiting for plant growth (Frost et al. 2005). Geider and La Roche (2002) argue that a realistic range of the critical N:P ratio lies between 15 to 30 (algae and cyanobacteria). For terrestrial plants at vegetation level, N:P ratios < 10 and > 20 correspond to nitrogen- and phosphorus-limited biomass production (Güsewell 2004). Vitousek and Howarth (1991) point out the difficulty or rather the impossibility in determining the degree of nutrient limitation on productivity. One reason is that not all plants change their C:N:P ratios in response to nutrient limitation in the same way (Sterner and Elser 2002). Species present in low nutrient sites may respond only little to nutrient addition, whereas others that require high levels of nutrients might have responded more had they been present (Howarth 1988, Vitousek and Howarth 1991). The authors therefore suggest conducting replicated nutrient addition experiments to evaluate the threshold that applies to each ecosystem.

3.4.2 Salt stress

The stoichiometry of plants growing in a salty environment differs from species exposed to low levels of salt. As this is especially important in the context of this thesis this paragraph addresses the influence of salt on the elemental compositions of plants.

Various studies verified the negative effect of salt on growth and reproduction of salt marsh plants (Jensen 1985, Lenssen et al. 1993, Lenssen et al. 1995, Egan and Ungar 2001, see Chapter 4.1.2). Accompanied by inundation frequency, salt concentration is regarded as primary stress factor in salt marshes (Rozema et al. 1985). The vacuole of terrestrial plant cells plays an important role in the homeostasis of the cell and is a major site for storing water, providing cell turgor pressure and permitting the cell to grow without investment in energy or nutrients (Wiebe 1978, Barkla and Pantoja 1996). High amounts of NaCl in the apoplast influence the equilibrium of the water and ion balances in the plant cell. The influx of Na^+- and Cl^--ions leads to an increase of Ca^{2+}-ions in the cytosol and a decrease of K^+. As a response the plant uses H^+-pumps to regain the membrane potential, which leads to decreased pH and changes in enzyme activity (Schulze et al. 2002). To prevent damage and to ensure further uptake of water and nutrients, Na^+- and Cl^--ions are compartmentalized in the vacuole, which would lead to unbalanced conditions between the vacuole and the cytoplasm (Gutknecht and Dainty 1969, Wyn Jones and Gorham 2002). To maintain osmotic equilibrium, plant cell synthesize 'compatible solutes' (Borowitzka 1981). Various molecules such as betaines (glycinebetaine), amiono acids (proline) and sugar alcohols (inositol, sorbitol) have been identified as compatible solutes in different species (Tab. 3-2).

Flowers and Colmer (2008) state that synthesis of different compatible solutes create different carbon and nitrogen costs. To evaluate the influence of these costs on the stoichiometry of salt marsh plants is one aim of this thesis.

Table 3-2: List of various molecules identified as compatible solutes in salt adapted plants.

Compatibel solute	Confirmed synthesis	Reference
Glycinebetaine	Present in numerous halophytes in a large number of families	Rhodes and Hanson (1993)
Proline	Present in halophytes from a wide range of families	Tipirdamaz et al. (2006)
Inositol	Present in Aizoaceae, Cyperaceae, Poaceae, Primulaceae	Gorham et al. (1980)
Pinitol	Chenopodiaceae, Plumbaginaceae	Gorham et al. (1981)
Sorbitol	Plantaginaceae	Koyro (2006)
Mannitol	Combretaceae	Stoop et al. (2006)

3.5 Allocation patterns

Allocation provides the basis for different strategies in ecology and is one of the basic concepts in modern ecology (Stearns 1992).

The fundamental processes in the life of a plant are growth and reproduction. A plant produces biomass and resources and allocates it to various organs and functions, for example reproduction (Bazzaz and Reekie 1985). By doing this, the resources allocated to one organ have become unavailable for another one, by that allocation implies trade-offs (Weiner 2004).

Two different approaches are presented in literature: 'partitioning' and 'allometry' (Weiner et al. 2009). Partitioning is a ratio-driven process (ratio-based perspective) for which it is assumed that at any point in time a plant has a given amount of resources and partitions it among its functions and organs as a response to variation in the environment (Klinkhamer et al. 1990, McConnaughay and Coleman 1999). For example, plants show larger root:shoot ratios as a response to limited water availability (Chapin et al. 1993) and when belowground resources are depleted, plants shift their allocation towards roots (Brouwer 1962b). The ratio-based perspective is size-independent, i.e. reproductive effort (reproductive biomass/total biomass) of a large plant may be similar to that of a small plant.

Allometry can be defined as the study of the relationship between size and shape and is a size-dependent approach (Small 1996, Weiner 2004). Allometric scaling relations are generally positive, e.g. leaf biomass increases with stem biomass (Enquist and Niklas 2002).

Similar to the study of plant allometry are biological scaling relationships, which are described by the equation

$$Y_0 = \beta Y_a^{\alpha} \qquad (1)$$

including Y_0 and Y_a as variables plotted on the ordinate and the abscissa axes, respectively, β as normalization constant and α as the scaling exponent (West et al. 1999). Isometric relationships are described by α = 1, i.e. the data fall into a straight line. When α ≠ 1, then equation (1) describes an anisometric relationship.

The exponential form can be transformed into a linear function

$$\log Y = \log \beta + \alpha \log Y_\alpha \qquad (2)$$

and β and α can be estimated from the intercept and slope of a linear regression analysis. The magnitude of β characterizes the rate of change of a biological variable subjected to a change of another variable and then reflects the geometric and dynamic constraints of the body (West et al. 1997).

It has been demonstrated that diverse organisms (unicellular algae, zooplankton, vascular plants and mammals) obey quarter-power scaling rules (Hemmingsen 1960, Peters 1983, Schmidt-Nielsen 1984, Niklas 1994b, Calder 1996). For example, annualized growth rates G scale as the ¾-power of body mass M ($G \propto M^{¾}$), plant body length L (i.e. cell length or plant height) scales as the ¼-power of M ($L \propto M^{1/4}$) and photosynthetic biomass M_p scales as the ¾-power of non-photosynthetic biomass M_n ($M_p \propto M_n^{¾}$) (Niklas and Enquist 2001).

The allometric approach has been applied to the metabolic theory of ecology (MTE), which posits that the metabolic rate of organisms is the fundamental biological rate that governs most observed patterns in ecology (Brown et al. 2004). However, MTE is explicitly concerned about temperature and body mass (Hillebrand et al. 2009), so that ecological stoichiometry (ES) emerged as approach to link the physiological processes within the plant cells to observed growth rates (Niklas 2006).

On the basis of selected elements, it was shown that nitrogen scales isometrically with carbon, phosphorus scales as the ¾ power with carbon, and so does nitrogen with phosphorus for a single plant (*Eranthis hyemalis*, Niklas and Cobb 2005). For 131 different angiosperms species Niklas et al. (2005) found that leaf nitrogen scales as ¾ power with leaf phosphorus. By relating scaling of nitrogen and phosphorus to growth rates, Niklas and Cobb (2005) concluded that growth depends on nitrogen and phosphorus allocation patterns ($G \propto N \propto P^{3/4}$; namely on rRNA and proteins) for both plants and animals.

4 Salt marshes

Coastal salt marshes may be defined as areas subjected to periodic flooding as a result of fluctuations in the level of the adjacent saline water body, which are vegetated by grasses, herbs or low shrubs (Adam 1990). Salt marshes form the upper part of the intertidal zone in tidal systems, i.e. the interface between land and sea. They extend from well below the mean high tide level up to the highest water mark (Esselink et al. 2009).

Salt marshes are exposed to constant changes in their environmental conditions and are controlled by physical, geomorphological and biological conditions, like tidal regime, sedimentation, erosion and progradation (Adam 1990, Bakker et al. 2005). Wadden Sea salt marshes make up 20% of the total area covered by salt marshes along the European Atlantic and Baltic coast. In Lower Saxony (NW-Germany, Fig. 4-1) mainland marshes extend over an area of 5,460 ha, whereas island marshes cover an area of 3,660 ha (data from 2004, Esselink et al. 2009).

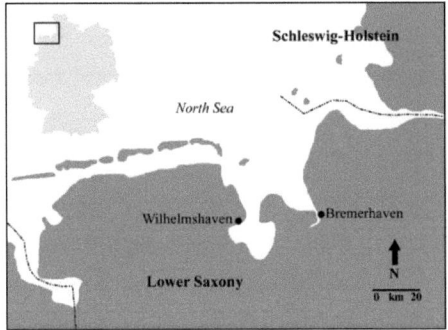

Figure 4-1: Overview of Germany (upper left) with part of Lower Saxony in corner. Magnification of detail giving the coastline of Lower Saxony and the Wadden Sea islands (dark grey shading). Cities of Wilhelmshaven and Bremerhaven give aid to orientation.

4.1 Abiotic conditions in salt marshes and their effects on plant performance

There are many environmental parameters influencing performance of salt marsh plants. For example, a high concentration of sulphide might itself be toxic for plants and further reduce the availability of Fe, Mn, Cu and Zn metals essential for a range of metabolic functions (Havill et al. 1985). However, environmental conditions are intermingled and might result from others, e.g. accumulation of reduced ions is a consequence of waterlogged soil conditions (Adam 1990). Thus the next paragraphs introduce three abiotic conditions accounted most important by literature, each of which affects plant performance in salt marshes by modifying the physical environment.

4.4.1 Inundation

Inundation is a dominant parameter of mechanical disturbance, which influences groundwater level and salinity, soil waterlogging, temperature, light and desiccation stress (Adam 1990, Van Diggelen 1991, Henley et al. 1992, Leuschner et al. 1998). It is assumed that inundation frequency and/or duration is connected with elevation and that the position of a plant along the elevation gradient is representative for the stress emerging from inundation (Chapman 1974). However, Bockelmann et al. (2002) showed that the geomorphology of the marsh (creek drainage system) and wind exposure may lead to uneven distribution of the incoming water, so that shore height is not a good indicator for inundation frequency.

Salt marsh species show different responses to inundation, however, it seems that that performance of upper marsh species is more affected than that of lower marsh species. Bouma et al. (2001) showed that the pioneer species *Spartina anglica* was best adapted to frequent flooding, second best the lower marsh species *Puccinellia maritima* and *Elymus athericus* (upper marsh) showed the least adaptations in terms of decrease in growth rate. Lenssen et al. (1995) demonstrated that biomass production of *Aster tripolium* (lower marsh) was stimulated by frequent flooding. Finally, Rozema et al. (1985) found mean relative growth rates of the upper marsh species *Elytrigia* (*Elymus*) *athericus*, *Festuca rubra* and *Juncus gerardii* were reduced when flooded by seawater, whereas lower marsh species were far less or not at all impeded in their growth.

4.4.2 Salinity

Rozema et al. (1985) called salinity the 'master' factor controlling zonation and succession in salt marshes and regarded inundation as additional operator to salinity. Growth of most halophytic species is inhibited by salt increments (despite *Salicornia*, *Suaeda* and *Atriplex*) and none shows maximal growth at seawater concentrations (Williams and Ungar 1972, Ungar 1991). Salinity tolerance changes through ontogenesis, in which seedlings are considered to be most sensitive to salt stress, as the young roots are exposed to higher salinities in the first centimeters of the soil surface (Ungar 1991).

Photosynthetic activity may be affected by salinity for several reasons. Stomatal conductance is decreased, which leads to reduced CO_2-uptake and impeded CO_2-conductance through the mesophyll cell (Guy and Reid 1986, Sekmen et al. 2007). A reduction of carboxylase content (e.g. Rubisco) decreases carbon-fixation (Flexas et al. 2004). Water deficits lead to a decrease in the amount of ATP and decreased regeneration of RuBP (Tezara et al. 1999) and high salt content inhibits repair of PSII.

Furthermore, salinity affects plant productivity by reducing the photosynthetic area by inhibiting cell division and cell expansion rates during leaf growth (Munns 2002) and decreased cell expansion in shoots (Munns and Tester 2008).

Jensen (1985) states that the salt tolerance depends on the amount of nitrogen in the medium. An increase in available nitrogen enables the plants to produce more nitrogen containing compatible osmotic solutes, which ensure enzyme activity and further uptake of water and nutrients, see chapter 3.4.2.

4.4.3 Nutrients

Nutrients like potassium and phosphorous are highest in the lower parts of a salt marsh and decrease with seaward distance (Ranwell 1964, Gray and Bunce 1972). Carbonate comes from the remains of mussels, which are washed into the marsh during flooding and its content decreases with elevation (Adam 1990). Olff et al. (1997) found a significant increase of nitrogen with increasing thickness of the silt layer. The nitrogen status of a marsh also depends on the input and leaching during flooding, the soil substrate (clay and sand content) and the mineralization rate (Rozema et al. 1985). Usually, mineralization outweighs humification in salt marshes (Beeftink 1966).

The availability of nitrogen is one of the major environmental factors limiting plant biomass production in saline habitats (Ungar 1991). Nitrogen is the limiting nutrient in temperate salt marshes and gets imported by groundwater and flood tides, sedimentation and fixation and exported by ebb tides and denitrification (Teal 1986, Kiehl et al. 1997). Only when nitrogen supply is sufficient, phosphorus gets limiting (Cargill and Jefferies 1984). Nitrogen uptake from the soil is inhibited by hypersaline conditions, low water-soil potentials, reducing environments and sulphide accumulation (Howes et al. 1986). The availability of nitrogen plays an important role in the salt stress regulation mechanism of salt marsh species, which has already been pointed out in various other studies (Rozema et al. 1985, Bakker et al. 1993, van Wijnen and Bakker 1997, see chapter 3.4.2). Salt marsh plants invest large portions of their nitrogen capital in compatible osmotic solutes, which enhances their salt tolerance and performance. Nitrogen availability controls processes like the productivity of primary and secondary producers and of decomposition rates of organic material (Teal 1986).

4.2 Species composition

Schematically, species composition in a salt marsh follows the elevation gradient, although it is a result of competition and environmental constraints (Bockelmann and Neuhaus 1999, Bockelmann et al. 2002).

Literature distinguishes three major zones (Pott 1995): pioneer zone, lower marsh and upper marsh (Fig. 4-2). The pioneer zone is situated approximately 40 to 25 cm below mean high tide (MHT) and shows the highest inundation frequency. The lower marsh can be found about 20 cm below to 30 cm above MHT and gets inundated about 250 times a year. Species of the upper marsh start to appear approximately 30 cm above MHT when inundation frequency ranges between 40 to 70 times a year (Pott 1995, Künnemann and Gad 1997). Although only infrequently inundated, the soil solution of the upper marsh may become hypersaline in periods of high insulation and temperature (De Leeuw et al. 1990).

Each salt marsh zone is inhabited by a distinct set of species, for which sequence of appearance is controlled by their ability to cope with abiotic and biotic conditions (e.g. salt tolerance and/or competitive exclusion by other species, Bockelmann and Neuhaus 1999).

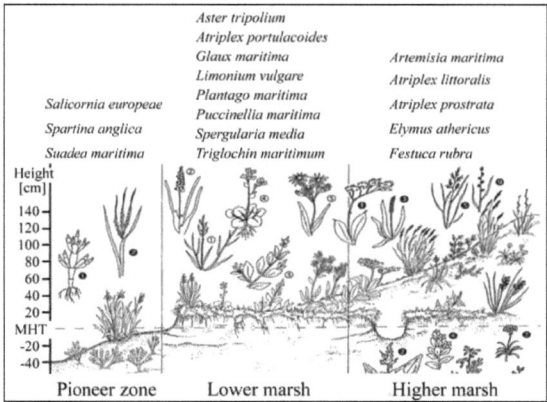

Figure 4-2: Composition of salt marsh species along the elevation gradient, with species names for each zone (pioneer zone, lower and upper marsh, adapted from Künnemann and Gad 1997).

4.3 Adaptations to salt stress

Despite the synthesis of compatible osmotic solutes (see Chapter 3.4.2), three general strategies have been detected in salt marsh plants, which are exclusion, dilution and morphological adaptations to osmotic stress (MAOS). Some species show a combination of several strategies, whereas others are restricted to a single adaptation, see Table 4-1.

4.3.1 Exclusion

Species might excrete salt ions through glands and bladders, reduce ion uptake through the root or accumulate ions in their tissue with subsequent shedding once concentration has exceeded a certain threshold (Kinzel 1982, Schirmer and Breckle 1982, Van Diggelen et al. 1986).

Excretion via glands is an active process which requires energy and as glands secrete a salt solution, it also involves loss of water, which might be disadvantageous under drought stress (Adam 1990). The salt solution is excreted to the outside of the leaf, where the water eventually evaporates and salt crystals remain, which are washed off by the rain (Adam 1990). Salt bladders occur in the family of the Chenopodiaceae (*Atriplex*-species) and are often two-celled structures which are positioned on the surface of the leaves. When the bladder is filled with salt it dies and bursts and the salt is lost from the plant (Packham and Willis 1997).

Reduction of ion uptake is achieved by *Puccinellia*-species by producing a second endodermal layer in the root. By this, the flow of solutes is forced into the symplast from which the ions have to cross the membrane in the root regulating their passage (Stelzer and Läuchli 1977).

In some species with no glands and little development of succulence, salt is accumulated in high concentrations in leaves and photosynthetic stems and is ultimately the determinant of the life of the plant organ (Kinzel 1982, Adam 1990).

4.3.2 Dilution

Some salt marsh plants 'dilute' the incoming ions in order to keep the ionic concentration in tolerable limits, which results in succulent growth (Kinzel 1982). Most of the succulent species are dicotyledons, whereas the majority of monocotyledons are never succulent (Adam 1990). The degree of succulence varies with external salinity and the dilution of cell sap is accomplished via enlargement of mesophyll cells. Increased succulence may result in reduced stomatal frequency, which reduces transpiration and uptake of further ions (Flowers 1975).

4.3.3 Morphological adaptations to osmotic stress (MAOS)

Some salt marsh plant show xeromorphic structures and features (Rozema et al. 1985). The transpiring area is reduced by rolling in the leaves, solar radiation is attenuated by sheathing of leaves and transpiration rate is diminished by production of epicuticular wax layers (Rozema et al. 1983). By reducing the transpiration rate, the uptake rate of saline water by the root is lowered.

Table 4-1: Names of species with abbreviation and adaptation to salt stress (MAOS: morphological adaptation to osmotic stress). Species names follow Flora Europaea+ in SynBioSys Species Checklist (2010). Abbr.: Abbreviation.

Species name	Abbr.	Adaptation to salt stress		
		Exclusion	Dilution	MAOS
Pioneer zone				
Salicornia europaea	Sal.eur	no	yes	no
Suaeda maritima	Sua.mar	no	yes	no
Spartina anglica	Spa.ang	yes	no	no
Lower salt marsh				
Aster tripolium	Ast.tri	yes	yes	no
Atriplex portulacoides	Atr.port	yes	no	no
Glaux maritima	Gla.mar	yes	yes	no
Limonium vulgare	Lim.vul	yes	no	no
Plantago maritima	Pla.mar	yes	no	no
Puccinellia maritima	Puc.mar	yes	no	no
Spergularia media	Spe.med	no	yes	no
Triglochin maritimum	Tri.mar	yes	yes	no
Upper salt marsh				
Artemisia maritima	Art.mar	no	no	yes
Atriplex littoralis	Atr.lit	yes	no	no
Atriplex prostrata	Atr.pro	yes	no	no
Elymus athericus	Ely.ath	no	no	yes
Festuca rubra	Fes.rub	no	no	yes

4.4 Ecosystem functions in salt marshes

Various ecosystem functions are covered by salt marshes and some of them are briefly summarized in the following.

Coastal marshes are among the world's most productive ecosystems (Howes et al. 1986, Bakker et al. 1993), however belowground production outweighs that of aboveground organs (Tyler 1971, Groenendijk and Vink-Lievaart 1987). This high productivity is the basis of other ecosystem functions, like provision of feeding ground or agricultural exploitation.

At first glance a salt marsh might look poor in species diversity. However, on European scale, nearly 200 of 1,068 plant species bound to the coastal zone are restricted to salt marshes (van der Maarel and van der Maarel-Versluys 1996). The highest diversity can be observed among the invertebrates; a considerable number of the 1,500 arthropod species inhabiting salt marshes are restricted to this habitat (Heydemann 1981). Also, salt marshes provide feeding, resting and breeding ground for many bird species (coastal waders, migratory birds (brent and barnacle goose), Koffijberg et al. 2003, Blew et al. 2005).

Decomposition rates are high in salt marshes, leading to rapid release and availability of carbon and nutrients essential for new growth and development of living organisms (Bakker et al. 1993). The rate of decomposition is most strongly controlled by tissue composition (mainly lignin and nitrogen content) and second by environmental conditions (Hemminga and Buth 1991).

The agricultural exploitation of salt marshes in Germany is only moderate and is restricted to cattle grazing and mowing, due to the foundation of the National Park Niedersächsisches Wattenmeer in 1985, including abandonment of extensive utilization (Bakker et al. 2005).

Tourism plays an important role in economy and the North Sea coast of Lower Saxony is a popular area attracting millions of visitors each year (Brandt and Wollesen 2009). In this context, the coastal marshes and the adjacent Wadden Sea area are a main attraction for recreation, tidal flat walking, water sports etc.

5 Study design and parameter-synopsis

5.1 Study sites

The study was conducted in salt marshes along the mainland coast of Lower Saxony, Germany and on the island of Mellum. Along the coast, data collection concentrated on three study areas, Leybucht, Norderland and Jade Bight (including Beckmannsfeld, Augustgroden and Dangast, Fig. 5-1).

Agricultural exploitation of mainland marshes included cattle grazing and mowing (once per year, start of June), Mellum is not influenced by land use. The study areas have a mean annual temperature of about 9°C and receive a precipitation from 770 to 830 mm per year (west to east, Deutscher Wetterdienst 2009).

The salt marshes along the mainland coast often developed though land reclamation for agricultural purposes (Pott 1995). Accompanied with the establishment of the National Park 'Niedersächsisches Wattenmeer' in 1985 anthropogenic influence has been reduced to a minimum (Esselink et al. 2009). Clayey silt, loamy sand and loamy silt are predominant in this area.

Opposite to that, the island of Mellum was hardly ever subject to anthropogenic influence (despite a small period during WWII, Kuhbier 1975) and as it belongs to the national park today, is preserved from human interference. The island developed from currents bringing sandy substrate from west to east along the coast of The Netherlands and Northwest Germany (Pott 1995). Although occasionally some soil types were identified as silty clay and clayey loam, the predominant part of Mellum consists of sandy substrate.

Data collection on Mellum was carried out by Hanna Timmermann and Janina Spalke. Field work on 57 plots in Norderland (fallow land) was accomplished by Sandra Andratschke.

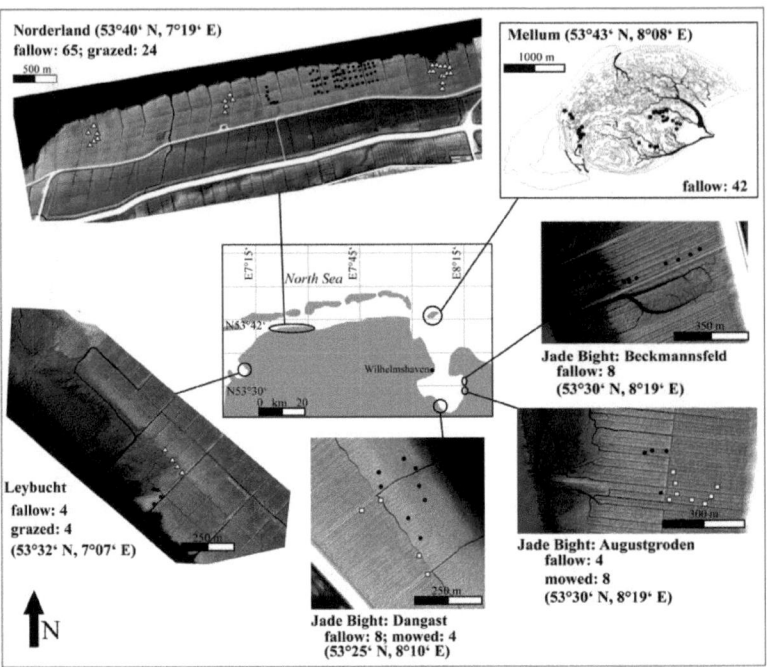

Figure 5-1: Overview of the study region (center; for position in Germany see Figure 4-1). Three study areas are located on the mainland (Norderland, Leybucht and Jade Bight (with Beckmannsfeld, Augustgroden and Dangast)) and one on the island of Mellum. Grid information is given for each area. Management regime is indicated for each plot by different symbols (fallow land (black circle), grazing (white triangle) and mowing (grey square)). Map of Mellum from Timmermann (2008).

5.2 Sampling design

Position of plots was determined by stratified random sampling, including two strata (Krebs 1989). To represent the species composition and environmental parameters along the elevation gradient in the dataset, the first stratum was elevation above sea level. To evaluate the influence of land use the second stratum included cattle grazed and mowed areas and fallow land, respectively. Random numbers were generated and provided distance measures from one corner of a plot to the corner of the next plot.

Each plot consisted of an area of 4 x 4 m, which was subdivided for the different data types (vegetation relevés, biomass removal, decomposition, soil samples and plant traits, see Fig. 5-2 left). In the centre of each plot a drainage pipe was installed in which groundwater level and salinity and inundation frequency were measured, the latter at a subset of 18 plots. In the study areas subject to land use, part of the plot was fenced ('exclosures', Fig. 5-2 right) to avoid disturbance (biomass removal by grazing and mowing, disturbance by trampling).

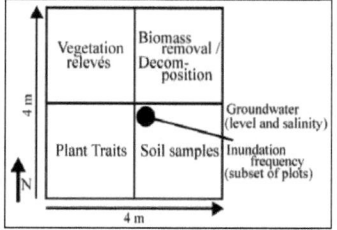 

Figure 5-2: Scheme of plot design (left), exclosure in areas with land use (right).

5.3 Species composition

Composition of vascular plant species was surveyed by frequency analysis in August 2007. A frame was laid out on the vegetation (1 m x 1 m), which was subdivided into 100 gridcells of 0.1 m x 0.1 m. In each gridcell, presence and absence of each species was recorded and summed up (Tremp 2005). Species names follow Flora Europaea+ (*Triglochin maritimum* and *Elymus athericus* follow German Turboveg flora) in SynBioSys Species Checklist (2010, see Table 4-1).

5.4 Abiotic parameters

Abiotic parameters focused on salinity and inundation frequency, nutrient availability and utilization by land use.

On each plot soil samples were taken to a depth of 30 cm due to upwelling groundwater. Bulk density was determined from 200 cm^3 of soil to transfer nutrient content to the area of measurement. The following parameters were measured from the soil samples: plant available potassium (Flame photometer, Egnér et al. 1960) and phosphorus (Continuous Flow Analyser (CFA), Murphy and Riley 1962), bulk density (Schlichting et al. 1995), sand content (Ad-Hoc-AG Boden 2005) and calcium carbonate (CaCO3, according to Scheibler in Schlichting et al. 1995).

Groundwater level and salinity was recorded biweekly from May to September 2007 at ebb tide and variation due to the tidal rhythm was subsequently adjusted by regression analysis. Salinity content of the groundwater was measured via a conductivity measurement device ('pH/Cond 340i' with measuring electrode 'Tetracon 325').

Inundation frequency was recorded by so-called 'divers'. These data loggers (ecoTech, Pegel-Datenlogger PDLA) gauged the water column in the drainage pipes at 18 plots distributed over the whole study region. Four additional loggers recorded the pressure of the surrounding air in each study area. Inundation frequency was calculated from the elevation of all plots relative to the water level measured by the data loggers.

To evaluate the influence of land use a utilization factor was calculated, which covered disturbance due to trampling and direct biomass removal (grazing, mowing). Biomass samples were taken in summer 2007 inside and outside of exclosures (Fig. 5-2 right). Offset of the values gave the direct biomass loss in percent. In the mown areas a utilization factor of 45% was used, assuming all belowground organs intact and remaining aboveground plant mass representing 5% of the former plant mass (i.e. intact mass being 55% and thus removal being 45%).

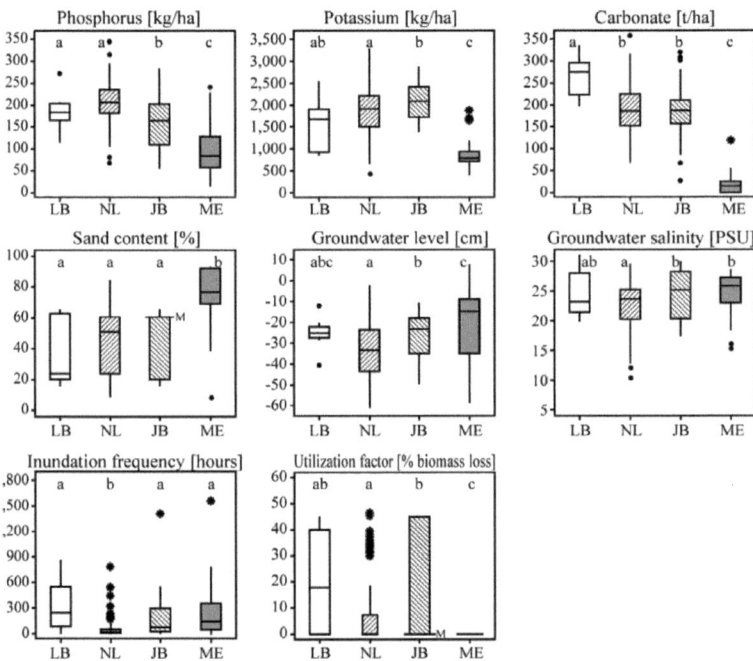

Figure 5-3: Abiotic parameters for the four study areas (LB: Leybucht, NL: Norderland, JB: Jade Bight, ME: Mellum). Circles indicate outside values (sample point > upper quartile + 1.5 times distance to quartile), asterisks far outside values (sample point > upper quartile + 3 times distance to quartile). Homogenous subgroups as results of t- and H-Test are shown by the use of the same letters (p-value > 0.05). 'M' next to boxplots indicates position of median when coinciding with upper and lower quartiles, respectively.

On the mainland phosphorus, potassium and carbonate values were higher, whereas sand content of the soil was lower than on the island (Fig. 5-3). Following Schlichting et al. (1995) mainland soils are high and very high in phosphorus and potassium, whereas soils on Mellum have moderate phosphorus and increased potassium supply.

Groundwater level and salinity and inundation frequency were similar between the study areas, as the plots were evenly distributed along the elevation gradient.

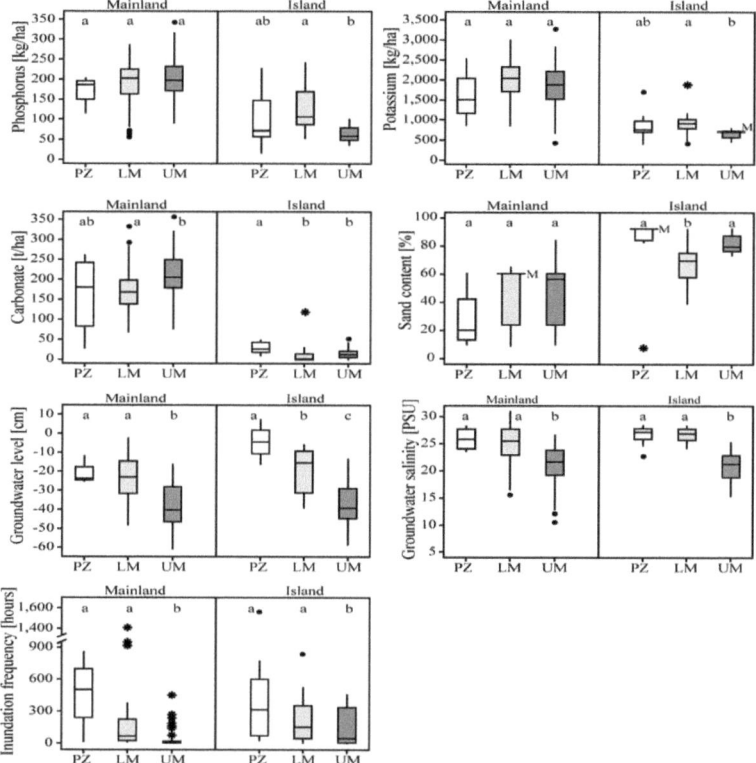

Figure 5-4: Abiotic parameters in the pioneer zone (PZ), lower marsh (LM) and upper marsh (UM) in mainland and island salt marshes. Circles indicate outside, asterisks far outside values (see Fig. 5-3). Homogenous subgroups as results of t- and H-Test are shown by the use of the same letters (p-value > 0.05). 'M' next to boxplots indicates position of median when coinciding with upper and lower quartiles, respectively.

On the mainland phosphorus, potassium and sand content of the soil do not differ between the different salt marsh zones (pioneer zone, lower and upper marsh, Fig. 5-4), carbonate shows an increase towards the upper marsh. Opposite to that, nutrient content in island marshes decrease towards the upper marsh and is generally lower than in mainland marshes. Environmental parameters related to the water regime (groundwater level and salinity and inundation frequency) decrease towards the upper marsh, both on mainland and island marshes.

Disturbance due to sedimentation or erosion was evaluated by so-called Sedimentation-Erosion-Bars (SEB's, Fig. 5-5). This method is based on differences in the distance of defined measuring points towards the ground level and was conducted in autumn 2006 and summer 2007.

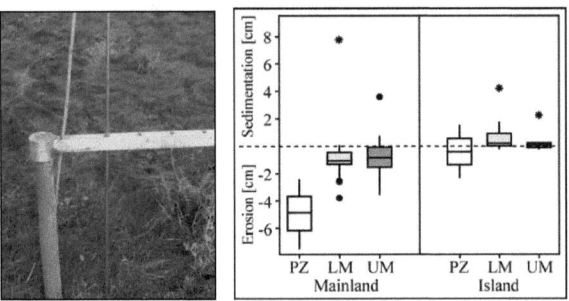

5-5: Sedimentation-Erosion-Bar (SEB, left) and results obtained by this method for pioneer zone (PZ), lower marsh (LM) and upper marsh (UM) for mainland and island marshes (right). Circles indicate outside, asterisks far outside values (see Fig. 5-3).

Erosion is highest in the pioneer zone and decreases towards the upper marsh both for mainland and island marshes. However, most mainland sites showed disturbance due to erosion, whereas on island sites both erosion and sedimentation was observed.

5.5 Plant functional traits

Most studies focusing on plant functional traits use so-called soft traits to address their hypotheses which are in the majority of cases restricted to aboveground plant features. This is mostly because hard traits are laborious and/or expensive (Weiher et al. 1999). Soft traits are then used as surrogates to express other functions, like SLA is used to draw conclusions about relative growth rate (Garnier et al. 1997, Violle et al. 2007).

The present study includes whole-plant traits, which allowed the unbiased examination of trait responses on environmental conditions and effects on ecosystem properties. However, collection and preparation of elemental-based traits was more laborious and time-consuming than of morphology-based traits. Traits were collected according to the recommendations of Knevel et al. (2005) and some traits were calculated from others (Table 5-1). Trait information is based on an average of nine individuals per species (each for mainland and island).

Table 5-1: List of plant traits with abbreviation and units. (* belowground organs refer to roots and rhizomes; † root includes roots and rhizomes; $ literature derived: Osmond et al. (1980), Rozema (1981), Kinzel (1982), Schirmer and Breckle (1982), Rozema (1985), van Diggelen et al. (1986).

Plant trait	Abbr.	Unit/Comment
Biomass related		
Aboveground dry mass (stems and leaves)	ADM	g
Belowground dry mass (roots and rhizomes)	BDM	g
Dry weight of diaspore	Diaspore	g
Dry weight of stems	Stem	g
Dry weight of leaves	Leaves	g
Dry weight of belowground organs*	Belowground	g
Reproductive Effort	RE	g (diaspore)/g (all)
Leaf:mass fraction	LMF	g (leaf)/g (all)
Stem:mass fraction	SMF	g (stem)/g (all)
Root:mass fraction†	RMF	g (root)/g (all)
Leaf related		
Specific Leaf Area	SLA	mm²/mg
Leaf Dry Matter Content	LDMC	mg/g
Stem related		
Specific Stem Density	SSD	mg/mm³
Stem Dry Matter Content	SDMC	mg/g
Leaf and stem related		
Leaf and stem properties	LSP	Composed of SLA, LDMC, SSD, SDMC
Height related		
Canopy height	Canopy	cm
Life history related		
Monocarpic plant	Mon	Yes/No
Clonal growth organ	CGO	Yes/No
Adapations to salt stress		
Exclusion of salt ions$	Exc	Yes/No
Dilution of cell sap$	Dil	Yes/No
Morphological Adaptations to Osmotic Stress$	MAOS	Yes/No
Related to elemental composition		
C:N ratio of whole plant	C:N	
Carbon, nitrogen and phosphorus content of		
Diaspores	C, N, P	%
Stems	C, N, P	%
Leaves	C, N, P	%
Belowground organs*	C, N, P	%

The allocation of carbon across the plant organs is highest in diaspores (mean of 43.2%, Appendix 1), second highest in stems (41.7%) and lowest in belowground organs and leaves (39.6 and 38.8 %). Considering differences between mainland and island species, carbon content is significantly higher in mainland species (Appendix 2) and increases along the elevation gradient.

Nitrogen and phosphorus contents are highest in diaspores (mean values of 2.4% and 0.3%), second highest in leaves (1.8% and 0.2%) and lowest in belowground organs (0.9% and 0.2%) and stems (0.9% and 0.1%, Appendix 3 and 5).

Opposite to carbon content, nitrogen and phosphorus contents of plant organs from mainland species are higher than those of island species and decrease towards the upper marsh (Appendix 4 and 6).

5.6 Ecosystem properties

The ecosystem properties of this study were chosen to refer to the nutrient and carbon cycle within salt marsh habitats and, by including species richness the role of trait expressions in relation of biodiversity is also highlighted.

Aboveground biomass (AGB) was sampled in August 2007 on an area of 0.5 m² on each plot, oven dried, weighted and subsequently extrapolated to 1m². Living biomass *(Aboveground biomass live, (AGB live))* was separated from dead plant parts *(Aboveground biomass dead, (AGB dead))*.

To evaluate *Aboveground Net Primary Productivity (ANPP)* additional biomass samples were taken in March 2007. ANPP is the difference between living biomass at the peak of the growing season and its beginning (August and March, De Leeuw et al. 1990), divided by the period of time (here, five months). Standing biomass samples were separated into living and dead parts, following Scurlock et al. (2002).

Decomposition rates were determined for 'Natural' and 'Standard' litter. Natural litter is the biomass produced at the plots, whereas standard litter is hey. Preparation followed Garnier et al. (2007). Decomposition is the amount of decomposed material in percent, and was determined after 6 and 12 months, respectively (Figure 5-6).

Species richness was based on the data of frequency analysis (see Chapter 5.3) and refers to the number of species found on each plot.

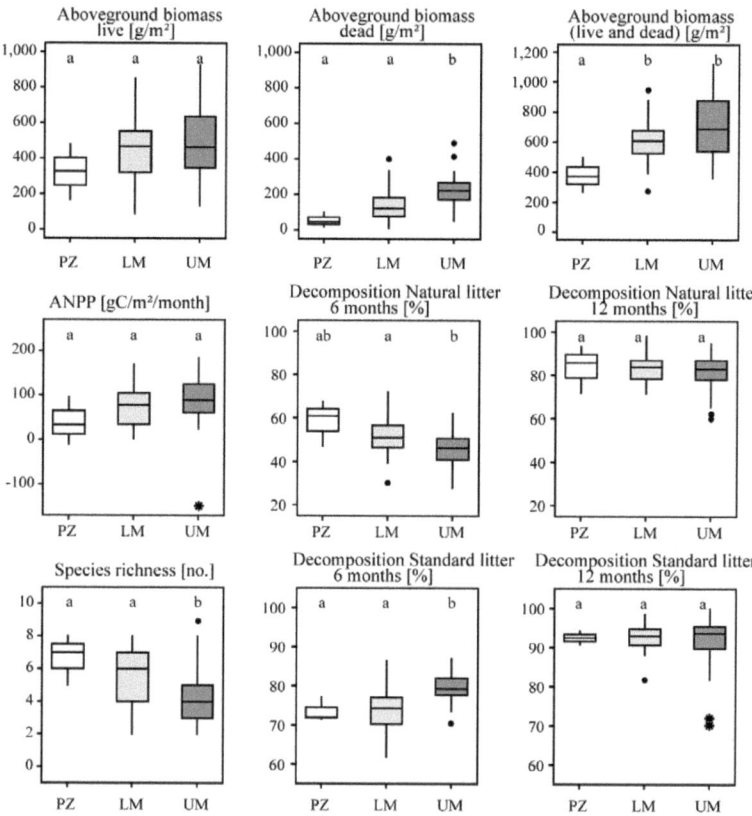

Figure 5-6: Ecosystem properties of this study (mainland marshes only). Values are ordered to refer to the zonation of the salt marsh (pioneer zone (PZ), lower marsh (LM) and upper marsh (UM)). Circles indicate outside, asterisks far outside values (see Fig. 5-3).

Whereas aboveground living biomass is similar between the pioneer zone, lower and upper marsh, the amount of dead and total biomass increase towards the upper marsh. Productivity is similar between the three salt marsh zones, as well as decomposition of natural and standard litter after 12 months. Species richness decreases towards the upper marsh, as well as decomposition of natural litter after 6 months, whereas decomposition of standard litter increases.

6 Trait-environmental concepts are not explicitly convertible to salt marshes

Vanessa Minden, Sandra Andratschke, Janina Spalke, Hanna Timmermann, Michael Kleyer, in preparation

Abstract

Salt marshes are exposed to high groundwater levels and salinity due to periodical inundation and are under constant influence of physical disturbance. Plants inhabiting salt marshes need to pursue different strategies to colonize this environment and hence, trait expressions should vary along the elevation gradient from pioneer zone to upper salt marsh. We verified various trait-environmental concepts for their applicability to different salt marshes zones, and found some are opposite to that of other terrestrial plants.

We used RLQ to analyse the joint structure between environmental conditions and species attributes by use of a species abundance table. Then we clustered species according to their position along the RLQ axes, and used the fourth-corner method to test for significance of the relationships between environmental parameters and species attributes.

The 'physiological-ecological-amplitude' concept by Scholten et al. (1987) could not be confirmed from the trait perspective, as the species of the 'stressful' lower marsh show trait values indicating higher competitive ability (canopy height and stem mass fraction) than those of the 'very benign' upper parts. Also, specific leaf area (SLA) and leaf dry matter content (LDMC) of salt marsh plants were more constrained by a salt-waterlogging gradient than by a nutrient gradient. This is opposite to the leaf economics spectrum (Wright et al. 2004), which describes a trade-off from fast growing species with the potential of quick return of investments of nutrients to species with long leaf life time and low rates of photosynthesis. Our results are consistent with Brouwer's (1962b) theory and Tilman's allocation model (1988).

Keywords: allocation, fourth-corner analysis, leaf economics spectrum, plant functional traits, RLQ-analysis

6.1 Introduction

Salt marsh plants are subject to constant changes of their habitat conditions. They are strongly controlled by geomorphological, physical and biological processes, such as sedimentation, tidal regime and wind-wave pattern (Bakker et al. 2005). Among the environmental conditions influencing establishment of salt marsh species, elevation in relation to tidal level is regarded the most important factor (Wolters et al. 2008). Both biotic and abiotic factors play a major role in determining the distribution of species along the elevation gradient. Duration of tidal flooding (Wiehe 1935), soil aeration (Armstrong et al. 1985), soil salinity and waterlogging (Cooper 1982, Snow and Vince 1984) are the main determinants of the seaward limit of salt marsh species, whereas the boundary towards the more benign upper part of the salt marsh is supposed to be the result of competition (Pielou and Routledge 1976, Pennings and Callaway 1992, Bockelmann and Neuhaus 1999, Davy et al. 2000). Species with high competitive ability monopolize benign habitats, and relegate inferior species to stressful habitats (Bertness 1991, Bertness and Leonard 1997, Engels and Jensen 2010). These interactions in salt marshes are represented in the 'physiological-ecological-amplitude' concept (Scholten et al. 1987), which has been discussed from species position on the elevation gradient or from experiments involving limited species numbers by many studies (Snow and Vince 1984, Scholten et al. 1987, Pennings and Callaway 1992, Bockelmann and Neuhaus 1999). According to Austin (1999), species niches in general are limited by physiological tolerances at the extremes of environmental gradients whereas competition controls the limits towards the centre of the gradient.

However, environmental stress in salt marshes can result from (i) a salinity gradient where stressful refers to high groundwater level and salinity and inundation frequency, or (ii) a nutrient gradient where stress refers to low levels of soil nutrients which are more often found on sandy marshes compared to clayey marshes. Also, nutrients like potassium and phosphorus are highest in the lower parts of a salt marsh and decrease with seaward distance (Ranwell 1964, Gray and Bunce 1972). It is not clear how these two gradients interact in distributing competitors and stress-tolerators on salt marshes.

In our study we ask whether the physiological-ecological-amplitude concept is supported by the functional trait composition of salt marsh plant communities, i.e. whether species occurring at the lower end of the elevation gradient show better adaptation to stress whereas species at the upper end invest more in traits conferring competitive effect. Furthermore, we ask if the trait-environment relationships of salt marshes are consistent with those found in other terrestrial habitats.

We use three sets of plant traits to indicate stress tolerance and competitive ability:

1. Salt stress adaption. This set of traits accounts for the ability of salt marsh species to cope with salt stress. These are (i) the exclusion of salt ions, (ii) dilution of cell sap and (iii) morphological adaptations to osmotic stress. Some salt marsh species exclude salt either by structural and functional adaptations that reduce salt uptake through the root or active excretion of ions via glands and bladders and/or constant renewal of their basal leaves (Kinzel 1982, Schirmer and Breckle 1982, Van Diggelen et al. 1986). Other species produce enlarged mesophyll cells by which the salt concentration of the cell sap is diluted and kept below toxic levels, which results in succulent growth (Kinzel 1982). The third strategy 'morphological adaptation to osmotic stress' refers primarily to species of the upper salt marsh (see Table 6-1), which adapt to osmotic stress-induced water deficits by rolling in their leaves or sheathe their leaves with hair in order to reduce solar radiation (Rozema et al. 1985).

2. Leaf economics spectrum. This set, known as components of the leaf economics spectrum (Wright et al. 2004), consists of traits relating the investment of biomass to the area of the photosynthetic active surface (e.g. specific leaf area) and the vertical extension of the structural tissue necessary to support carbon gain (e.g. canopy height and stem mass fraction). These traits describe a gradient from slow growing species, retention of nutrients and biomass and stress tolerance to potentially fast growing species, high acquisition of resources, as well as strong light interception and shading (Wright et al. 2004). Across many terrestrial habitats, they are accounted as responsive to disturbance intensity and soil resource availability (i.e. high SLA at high levels of resources, Poorter and de Jong 1999, Kühner and Kleyer 2008).

3. Plant allocation pattern. This set of traits describes the plant allocation pattern in response to the resources supplied by the environment. Resource allocation patterns have usually been represented in terms of ratios of biomass of plant organs. The theory of functional equilibrium of Brouwer (1962b) states that plants shift their allocation toward shoots when the carbon gain of the shoot is abated by low levels of aboveground resources, such as light and CO_2. On the other hand, when levels of belowground resources are impaired, such as nutrient availability and water, plants shift their allocation towards roots, which might lead to higher belowground competition (Berendse and Möller 2009). This creates a trade-off between below- and aboveground plant organs involved in capturing different resources and may also determine investment in reproduction (Obeso 2002). Rather than using root:shoot ratios, we follow the proposal of Körner (1994) and Poorter and Nagel (2000) to express the biomass of leaves, stems and roots as fractions of the total plant biomass because the combination of stems and leaves into shoots neglects the different functions of stems

and leaves. Additionally, we use reproductive effort which is diaspore mass as fraction of the total plant biomass.

Salt marshes differ in more than just elevation and inundation frequency. Nutrient regimes differ strongly depending on the amount of clay in the soil (Olff et al. 1997), groundwater tables determine soil aeration (Armstrong et al. 1985), whereas anthropogenic land use and natural sedimentation or erosion provide physical disturbance for the plants (van Eerdt 1985). Rarely have trait-environment studies on salt marshes taken so many factors into account. In this study, we quantify the response of the functional composition of salt marsh species to inundation frequency, groundwater levels, salinity of the groundwater, soil nutrients and utilization (i.e. biomass removal via cattle grazing or mowing). To disentangle responses to salt stress and nutrients, we sampled the inundation gradient on clay-rich salt marshes of the mainland coast and on sandy salt marshes of the island of Mellum in NW-Germany.

Using the three set of traits from above we focus on the question if they respond to the salt marsh environment and if the combination of these traits is indicative of environmental stress and competition. We expect that (i) plants should shift from exclusion and dilution of excessive salt in frequently inundated sites with high groundwater level and salinity to morphological adaptation to osmotic stress by reducing the transpiration rate of the aboveground tissue in infrequently inundated sites (Rozema et al. 1985), (ii) SLA (specific leaf area) and canopy height will increase with increasing nutrients and utilization as well as decreasing inundation frequency, ground water levels and salinity and accordingly we expect (iii) leaf and stem dry matter content to decrease (expectation (ii) and (iii) both refer to the physiological-ecological-amplitude concept by Scholten et al. (1987) and to the leaf economics spectrum by Wright et al. (2004). Revolving Brouwer's (1962b) theory of functional equilibrium, we expect that (iv) high nutrient availability negatively affects root biomass and positively affects stem biomass.

6.2 Materials and methods

Study area

The study took place in mainland salt marshes along the coast of Lower Saxony and in salt marshes on the island of Mellum, Germany (Fig. 6-1). The study region has a mean annual temperature of about 9°C and receives a precipitation from 770 mm to 830 mm per year (west to east, Deutscher Wetterdienst 2009).

Mainland salt marshes often developed through land reclamation for agricultural use and seawall protection (Pott 1995). Nowadays all salt marshes in the Wadden Sea are under national nature protection (Bakker *et al.* 2005). Clayey silt, loamy sand and loamy silt are predominant soil substrates in the study area. Along the mainland coast, elevation ranges from 0.2 m to 1.1 m above Mean High Tide (MHT). On the mainland, three different study areas were selected, these were Leybucht (LB, 53°32'N, 7°07'E, 8 survey plots), Norderland (NL, 53°40'N, 7°19'E, 89 plots) and Jade Bight (JB, 53°26'N, 8°09'E, 32 plots). Parts of the areas were subjected to cattle grazing and mowing, whereas the major part was not exploited.

The island of Mellum (I, 53°43'N, 8°08'E, 42 plots) developed from barrier walls and consists predominantly of sandy substrate (Reineck 1987). Today it is part of the National Park 'Niedersächsisches Wattenmeer' and is preserved from human interference. In the salt marshes of Mellum, elevation ranges from 0.0 m to 1.0 m above MHT.

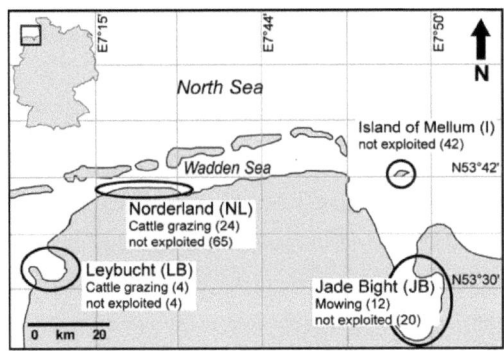

Figure 6-1: Position of study region in Germany (left upper corner) and location of study areas along the northwest coast of Germany with three study areas (circles) along the mainland coast and one on the island of Mellum.

Almost all abiotic factors differed significantly between mainland and island salt marshes: phosphorous, potassium, calcium carbonate ($CaCO_3$), sand content of the soil, salinity of the groundwater, and average groundwater level (Oneway ANOVA, p-value 0.05). In contrast to the mainland, there are no grazing animals on Mellum, apart from staging geese. Thus, these two types of salt marshes were considered different habitats for this study.

For vegetation three main salt marsh zones can be distinguished (Pott 1995): (a) the pioneer zone approximately 40 cm below mean high tide (MHT), (b) the lower marsh approximately 20 cm below to 30 cm above MHT and (c) the upper salt marsh positioned 30 cm above MHT and more (Künnemann and Gad 1997). Species composition differed in each salt marsh zone (Tab. 6-1, Pott 1995).

Table 6-1: Species names of pioneer zone, lower and upper salt marsh, with abbreviation in brackets, used for RLQ-table L. Species names follow Flora Europaea+ (*Triglochin maritimum* and *Elymus athericus* follow German Turboveg flora) in SynBioSys Species Checklist (2010). Trait samples were taken (x) or not taken (-) for species of the mainland and the island, depending on their occurrences in the plots. Number of measured individuals given in brackets. Different adaptations to salt stress are expressed by 'yes' and 'no'. MAOS: morphological adaptation to osmotic stress.

	Sampled for		Adaptation to salt stress		
Species name (Abbreviation)	Mainland	Island (Mellum)	Excl.	Dil.	MAOS
Pioneer zone					
Salicornia europaea (Sal.eur)	x (8)	x (10)	no	yes	no
Suaeda maritima (Sua.mar)	x (8)	x (10)	no	yes	no
Spartina anglica (Spa.ang)	x (8)	x (10)	yes	no	no
No. of measured individuals	24	30			
Lower salt marsh					
Aster tripolium (Ast.tri)	x (8)	x (10)	yes	yes	no
Atriplex portulacoides (Atr.port)	x (8)	x (10)	yes	no	no
Glaux maritima (Gla.mar)	-	x (10)	yes	yes	no
Limonium vulgare (Lim.vul)	x (8)	x (10)	yes	no	no
Plantago maritima (Pla.mar)	x (8)	x (10)	yes	no	no
Puccinellia maritima (Puc.mar)	x (8)	x (10)	yes	no	no
Spergularia media (Spe.mar)	-	x (10)	no	yes	no
Triglochin maritimum (Tri.mar)	x (8)	x (10)	yes	yes	no
No. of measured individuals	48	80			
Upper salt marsh					
Artemisia maritima (Atr.mar)	x (8)	-	no	no	yes
Atriplex littoralis (Atr.lit)	x (8)	-	yes	no	no
Atriplex prostrata (Atr.pro)	x (8)	-	yes	no	no
Elymus athericus (Ely.ath)	x (8)	x (10)	no	no	yes
Festuca rubra (Fes.rub)	x (8)	x (10)	no	no	yes
No. of measured individuals	40	20			

Sampling design

On the mainland, 129 plots were established and 42 plots on the island of Mellum (4 m x 4 m each). Plots were chosen by random stratified sampling. The strata were height above sea level (0.0 m to 1.5 m above sea level, covering the elevation gradient) and disturbance (mowing, grazing and no exploitation, respectively, see Krebs 1989). Random numbers were generated and provided distance measures from one corner of a plot to the corner of the next plot.

Field and laboratory measurements

Species composition and abundance was recorded by frequency analysis at each plot. In a subplot facing southwest a 1 m x 1 m frame was laid out, which was subdivided into 100 gridcells of 0.1 m x 0.1 m. In each of the gridcells presence or absence of each species was recorded and summed up, i.e. a species which occurred in each gridcell was attributed the number 100 (Tremp 2005).

To measure soil fertility, soil samples were taken at each plot. Due to high groundwater levels, it was not possible to sample below 30 cm at many plots. Therefore, all soils were sampled to a depth of 30 cm, to make plots comparable. Bulk density was evaluated from 200 cm^3 of soil, see Schlichting et al. (1995). Subsequently, soil samples were air dried, sieved through a 2 mm sieve and analysed for sand content (Tab. 6-2, Ad-Hoc-AG Boden 2005). Calcium carbonate ($CaCO_3$) was determined by adding 10 ml hydrochloric acid (dilution 1:3) to a 10 g soil sample and by measuring the carbon dioxide produced (gasometric technique, according to Scheibler in Schlichting et al. 1995). Plant available potassium and phosphorous were extracted with ammoniumlactate-acetic acid at pH 3 following Egnér et al. (1960) and analysed by AAS (Atomic Adsorption Spectroscopy) and CFA (Continous Flow Analyser, Murphy and Riley 1962), respectively.

At each plot, a drainage pipe (6.5 cm diameter, 171 in total) was installed 80 cm vertically in the ground. In these pipes the groundwater level was recorded biweekly from May to September 2007 at ebb tide, as well as the salinity content of the groundwater via a conductivity measurement device ('pH/Cond 340i' with measuring electrode 'Tetracon 325'). Variation due to the tidal rhythm was adjusted by regression analysis.

To record inundation frequencies data loggers ("diver", ecoTech, Pegel-Datenlogger PDLA) recorded the water column in 18 drainage pipes each hour from May to September 2007. In each of the study areas (Leybucht, Norderland, Jade Bight and Mellum) one additional data logger was placed nearby the study plots to record the pressure of the surrounding air, which was needed to calculate the relative pressure of water accumulating in the pipes. Inundation frequency was calculated from the elevation of all plots relative to the water level measured by the data loggers.

Some mainland marshes were mown or grazed by cattle. To assess the influence of management on the vegetation, a utilization factor was calculated. In the cattle grazed areas, plots were fenced to prevent grazing. In summer 2007 biomass was cut on an area of 0.5 m^2 on each plot, inside and outside the fence, respectively. Offset of the values gave the influence of cattle grazing in percent. The utilization factor represents the sum of (i) the herbage consumed by the cattle, and (ii) losses of herbage due to trampling (Bakker 1989). For mown areas we used a utilization value of 45 %, i.e. percentage of removed biomass by mowing. It was assumed that belowground plant organs were not affected by mowing or cattle grazing. Aboveground cutting left short stubbles of the vegetation which added up to 55 % of intact vegetation after mowing.

Plant functional traits

Traits were collected for different species of the mainland and the island (Table 6-1 and 6-2). Trait information was based on measurements of 242 individuals (8 and 10 individuals for every mainland and island species, respectively).

Plants were dug out, roots and rhizomes were cleaned of soil material by rinsing off the soil substrate and roots of different individuals were carefully separated using tweezers. Plant material was subsequently oven dried at 70 °C for 72 hours. Leaves, stems, diaspores, roots and rhizomes were weighted after drying. Leaf mass fraction (LMF) was calculated as proportion of leaves relative to the biomass of the whole plant (g (leaf) / g (all)), as were stem mass fraction (SMF), root mass fraction (RMF) and reproductive effort (RE). As it was difficult to distinguish between root and rhizomes, the term 'root' includes both organs. Canopy height is defined as the distance between the highest photosynthetic tissue and the base of the plant (Weiher et al. 1999). SLA (specific leaf area) was calculated as the ratio of leaf area to leaf dry mass (mm^2/mg). For species like *Salicornia europaea* the top 2 cm of a twig was used as leaf analogue, as recommended by Knevel et al. (2005). LDMC (leaf dry matter content) is the ratio of dry leaf mass to fresh leaf mass (mg/g), the same counts for SDMC (stem dry matter content). Calculation of SLA, LDMC and SDMC followed Knevel et al. (2005). The plant traits monocarpic life span (Mon) and clonal growth organs (CGO) were observed in the field. Morphological adaptations to osmotic stress (MAOS), exclusion of salt ions (Exc) and dilution of cell sap (Dil) were literature derived plant traits, see Table 6-2.

Statistical analysis

Groundwater level

Groundwater level was recorded during low tide and thus lacked information about tidal variation. To adjust this, a regression was conducted with paired values of the hourly data produced by the 18 loggers and the biweekly data of the groundwater levels of all plots. Mean values of both groundwater and logger data over the measurement period were generated and used for linear regression analysis. The regression function was used to adjust values of mean groundwater level of all other plots to include information about high tide.

RLQ-analysis

Aim of our analysis was to relate species traits to environmental conditions, considering the abundances of species in the plots, for which RLQ-analysis is an adequate method (Dolédec et al. 1996, Legendre et al. 1997, Ribera et al. 2001, Dray et al. 2003, Choler 2005).

The analysis investigates the joint structure among three tables, i.e. R-table (containing environmental variables), Q-table (species traits), and species abundance table (L-table, Dolédec et al. 1996, Dray et al. 2002). Table L serves as a link between R and Q, and measures the intensity of the relationship between them. Before the actual analysis, three separate analyses were accomplished. A correspondence analysis (CA) is applied on the L-table. The CA gives the optimal correlations between the study sites and the species scores. Ordination of table R and L was done by principal component analysis (PCA). Column weights of table L were used for ordination of table Q, also by PCA. Inundation (environmental variable) and diaspore mass (trait) were transformed (log 10) prior to the analysis to achieve normal distribution (Leyer and Wesche 2007).

Subsequently, RLQ performed a co-inertia analysis on the cross-matrix of R, L, and Q. This analysis maximizes the covariance between the study site scores constrained by the environmental variables of table R and the species scores constrained by the traits of table Q. As a result, the best joint combination of the ordination of sites by their environmental characteristics, the ordination of species by their attributes (traits), and the simultaneous ordination of species and sites is calculated (Thuiller et al. 2006).

Following RLQ, a clustering was accomplished using hierarchical clustering of species scores on the first two RLQ axes (Ward's method). Subsequently, the optimal number of groups was determined using Calinski criteria (Calinski and Harabasz 1974). These clusters show the distribution of functional groups in the trait-environment space.

Fourth-corner analysis

To test for significant relationships between trait characteristics and environmental conditions at the plots we used the fourth-corner analysis, which was first implemented by Legendre (1997). We used the version by Dray and Legendre (2008), which allows the use of quantitative data in the species data matrix. The analysis tests for significance of the link between all combinations of environmental variables and species traits. The quantitative variables were standardized and qualitative variables were coded using dummy variables (Dray and Legendre 2008). Testing of the significance of the relationship between trait attributes and environmental conditions was done by permutation procedure. We chose permutation model 1, which was the most appropriate due to strong environmental gradients in our study areas (Dray and Legendre 2008). The null hypothesis (H_0) of this model states the random distribution of species with respect to site characteristics. The alternative hypothesis (H_1) suggests that the environmental preference of a species defines its distribution.

All analyses were conducted using ade4 package in R (Thioulouse et al. 1997, The R Foundation for Statistical Computing 2008). We used 'fourthcorner', with 999 permutations.

6.3 Results

RLQ- and fourth-corner analysis

The first two axes of the RLQ-analysis explained 97% of the total variation (71% and 26%, respectively), indicating that the environmental conditions can be summarized on two gradients. Although the first RLQ axis showed high correlations to all environmental parameters (Table 6-2), Fig. 6-2a reveals an orthogonal arrangement of two major sets of environmental variables which are rotated diagonally to the RLQ axes. The first set of variables, i.e. potassium, phosphorus, $CaCO_3$, sand content and utilization, spanned a 'nutrient' gradient from nutrient rich sites on the mainland to nutrient poor sites found on the island (Fig. 6-2c). A second 'salt-waterlogging stress' gradient, uncorrelated to the first gradient, was determined by inundation frequency, groundwater salinity and groundwater level. It separated highly inundated sites with high levels of salty groundwater from infrequently inundated sites with low groundwater levels. Highly inundated sites were found in the lower parts of the mainland salt marshes and the island marshes, whereas the upper marsh of the both mainland and island were only infrequently inundated (Fig. 6-2d). The ordination space of the RLQ-analysis revealed four different zones along the gradients. First, a nutrient rich zone with high influence of inundation and groundwater level and salinity (upper left corner, Fig 6-2a), second and opposite to that, a nutrient depleted zone with low inundation frequency and groundwater level and salinity. The third zone is characterized by high influence of the salt-waterlogging stress gradient, but low availability of nutrients (upper right corner, Fig 6-2a), whereas the fourth zone shows high nutrient availability and low influence of the salt-waterlogging stress gradient (opposite to zone three).

Canopy height (Cano), stem mass fraction (SMF) and reproductive effort (RE) increased and leaf mass fraction (LMF) decreased with the nutrient gradient (Fig. 6-2b), which was also confirmed by the results of the fourth-corner analysis (Table 6-3). On the other hand, traits associated with salt adaption and leaf economics followed the salt-waterlogging stress gradient. Excretion of excessive salt (Exc) and dilution of salt (Dil) increased with salt-waterlogging stress whereas morphological adaptations to osmotic stress (MAOS), leaf and stem dry matter content (LDMC and SDMC) decreased. Increase in specific leaf area (SLA) was only moderately related to increasing nutrients and salt-waterlogging stress whereas root mass fraction (RMF) increased with increasing salt-waterlogging stress and decreasing nutrients.

Monocarpic life span (Mon), clonal growth organs (CGO) and diaspore mass (Dia) only showed weak relationships to the environmental variables, probably because they are distributed more or less equally along the gradient, so that an assignment to a specific end of a gradient is not possible. On the other hand, utilization as an indicator of disturbance only showed weak relationships to the traits. The stocking with cattle on the studied salt marshes was relatively low (animal units 0.5 to 1.5 animals per hectare) and others parts of the areas were only mown once per year. It appeared disturbance by utilization was too low to influence trait expressions strongly. These traits and utilization are both omitted from further discussion.

Table 6-2: Environmental variables recorded on study plot, used for RLQ-table R and correlation with first two RLQ axes; trait information for each species, used for RLQ-table Q and correlation with first two RLQ axes. Source for MAOS, Dil and Exc: Osmond et al. (1980), Rozema (1981), Kinzel (1982), Schirmer and Breckle (1982), Rozema (1985), van Diggelen et al. (1986). [a]Root refers to belowground organs, i.e. roots and rhizomes. RLQ = RLQ axis

Environmental variable	Abbr.	Unit	RLQ 1	RLQ 2
Sand content	Sand	%	**0.57**	-0.42
Calcium carbonate	CaCO$_3$	t/ha	**-0.88**	0.31
Potassium	Pot	kg/ha	**-0.57**	**0.63**
Phosphorous	Phos	kg/ha	**-0.69**	0.34
Mean level of groundwater	GW.mean	cm	**0.66**	**0.51**
Mean salinity of groundwater	GW.sal	PSU	**0.51**	**0.58**
Inundation frequency	Inun	hours	0.45	**0.55**
Utilization	Util	% biomass loss	-0.47	0.28
Trait	**Abbr.**	**Unit, Attributes**	**RLQ 1**	**RLQ 2**
Canopy height	Cano	cm	**-0.65**	**0.67**
Leaf:mass fraction	LMF	g (leaf)/g (all)	**0.57**	-0.47
Reproductive effort	RE	g (diaspore)/g(all)	**-0.63**	0.40
Root:mass fraction[a]	RMF	g (root)/g (all)	**0.77**	-0.09
Stem:mass fraction	SMF	g (stem)/g (all)	**-0.77**	0.37
Stem dry matter content	SDMC	mg g^{-1}	-0.07	**-0.86**
Leaf dry matter content	LDMC	mg g^{-1}	-0.48	**-0.75**
Morphological adaptations to osmotic stress	MAOS	1 – Yes/ 0 - No	**-0.67**	-0.28
Exclusion of salt ions	Exc	1 – Yes/ 0 - No	0.20	**0.61**
Dilution of cell sap	Dil	1 – Yes/ 0 - No	0.37	**0.63**
Specific leaf area	SLA	mm^{2}mg^{-1}	-0.13	**0.73**
Monocarpic plant	Mon	1 – Yes/ 0 - No	-0.07	**0.78**
Clonal growth organ	CGO	1 – Yes/ 0 - No	0.03	-0.31
Diaspore mass	Dia	g	-0.23	-0.17

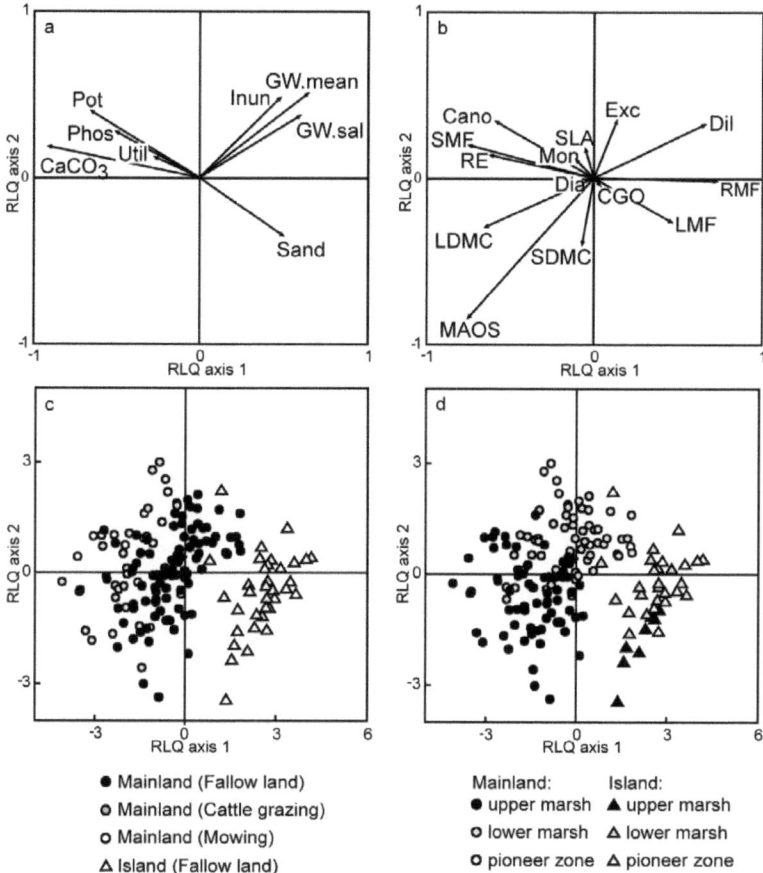

Figure 6-2: Ordination diagrams of the first two axes of the RLQ-analysis displaying the environmental variables (A) and plant traits (B; abbreviations explained in Table 6-2). Partitioning of the study sites is done by utilization (C) and elevation (D); see also symbols underneath.

Table 6-3: Results of the fourth-corner analysis, in which environmental variables are related to traits. All relationships are at p≤ 0.05 and are either positive or negative. Number in cells are correlation coefficients (r), empty cells indicate no significant correlation. For abbreviations see Table 6-2.

	Phos	Sand	CaCO$_3$	Pot	Util	GW.m.	GW.s.	Inun
Cano	0.27	-0.29	0.39	0.35	0.19	-0.08	-0.09	
LMF	-0.18	0.20	-0.31	-0.31	-0.19	0.08	0.08	
RE	0.22	-0.20	0.36	0.30	0.20	-0.16	-0.20	
RMF	-0.23	0.25	-0.40	-0.29		0.26	0.24	0.24
SMF	0.29	-0.32	0.44	0.39		-0.19	-0.16	-0.17
SDMC	-0.11	0.14		-0.08		-0.26	-0.14	-0.22
LDMC	0.15	-0.10	0.31	0.12		-0.39	-0.32	-0.37
MAOS			0.16			-0.45	-0.41	-0.35
Exc		-0.10		0.15		0.28	0.27	0.22
Dil			-0.13	-0.11		0.21	0.14	0.18
SLA		-0.07		0.10	0.08	0.08		0.09
Mon	0.07	-0.07						
CGO	0.07	0.07						
Dia					-0.12			

Cluster analysis

The cluster analysis yielded five stable functional groups showing how multiple trait expressions combine in species groups (Fig. 6-3). All clusters were clearly separated in the trait-environment space. Clusters E, A, and B were orientated along the increasing nutrient gradient, whereas clusters C and D were arranged along the increasing salt-waterlogging stress gradient. On the other hand, cluster D could also be viewed on the nutrient gradient, because it is positioned close to the center of both gradients. Clusters A and B contained only species from the mainland, whereas their island counterparts were placed in clusters D and E, indicating that these species respond in a plastic way to the nutrient gradient. Opposite to that, trait-plasticity of *Festuca rubra* and *Elymus athericus* was low, because the respective species from mainland and island were grouped together in cluster C.

Cluster A combined mainland species of the pioneer zone and the lower marsh. Cluster B contained only two species, one of the lower and one of the upper salt marsh. Species of both clusters primarily exclude or dilute salt ions, and showed high SMF and RE, low LMF and RMF and intermediate LDMC and SDMC (Fig. 6-4). Species of cluster B had the highest canopy height and SLA compared to all other clusters. Species in clusters D and E exclude or dilute salt ions. However, contrary to clusters A and B, they showed high LMF and RMF, low canopy height, RE, and LDMC, and intermediate SDMC. Cluster D had higher SLA than cluster E. Species from cluster C were characterized by morphological adaptations to osmotic stress. In addition they showed high RE, LDMC, and SDMC, low LMF and intermediate canopy height, SLA, RMF, and SMF.

Altogether, the clusters varied most significantly in carbon allocation traits and canopy height. Clusters A and B comprised the largest plants with highest investments in stems and reproduction and occurred at nutrient rich sites. Species of clusters D and E were small with strong allocation to roots and leaves and were found on nutrient poor sites (compare with Fig. 6-2d), with cluster D on sites with higher salt-waterlogging stress. Species of cluster C took an intermediate position in allocation patterns but showed higher LDMC and SDMC than all other clusters. They occurred on sites with low salt-waterlogging stress and intermediate to high nutrient levels.

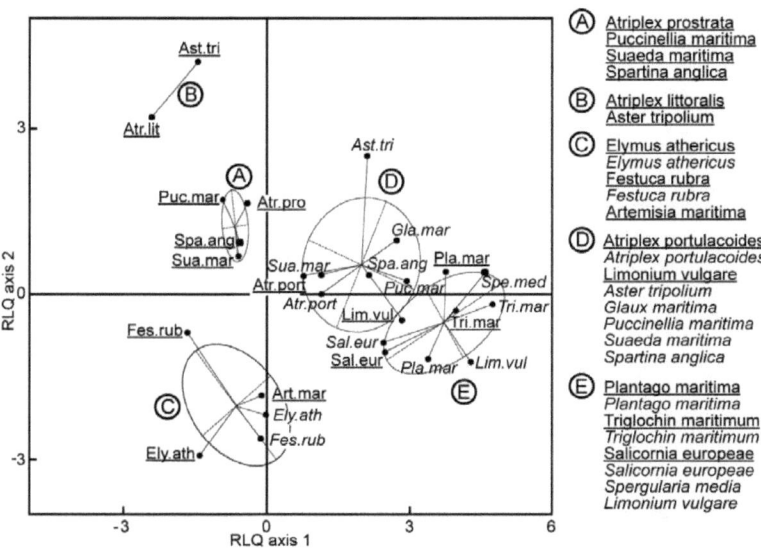

Figure 6-3: Species cluster (A-E) along the first two RLQ axes, with separation of species in Mainland (bold) and Island (italic) on right side.

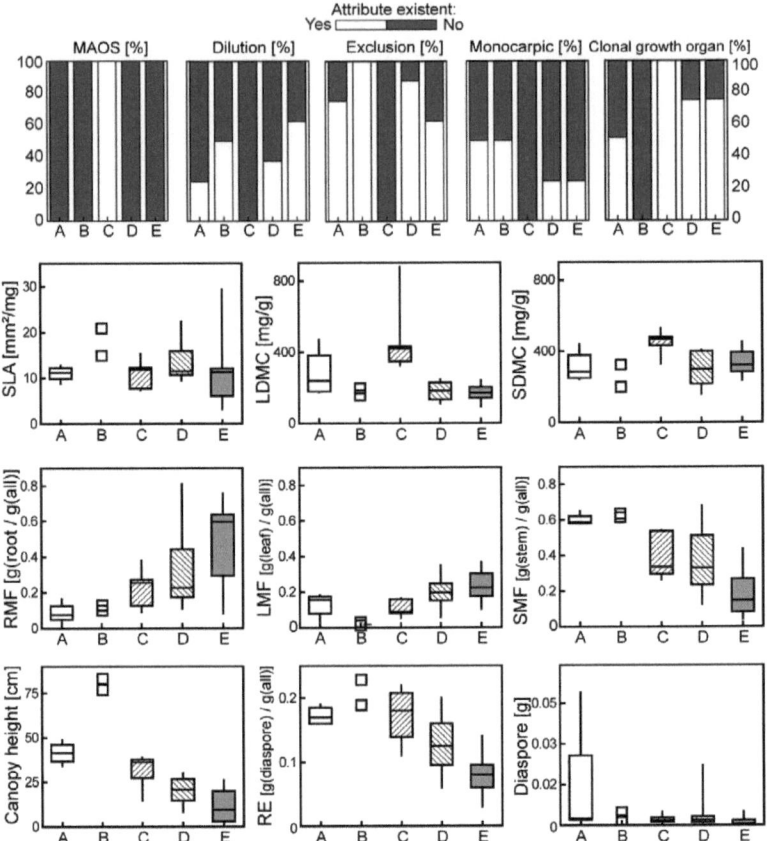

Figure 6-4: Trait attributes of each cluster (A to F). The vertical bar graphs show the nominal variables with scaling 'Yes' or 'No' in percent for each cluster (MAOS: morphological adaptations to osmotic stress). Boxplots show quantitative variables with their ranges in each cluster.

6.4 Discussion

Rarely have studies investigated trait responses to the interaction of a resource gradient with a direct environmental factor that cannot be consumed, such as salt (Austin and Smith 1989). In our study, this interaction produced trait-environment linkages that did only partly confirm to our hypotheses. In particular, traits associated with the leaf economics spectrum showed different responses than those of terrestrial habitats, whereas vegetative biomass allocation ratios followed the predictions by Brouwer (1962b) or Tilman (1988).

Submergence due to inundation, waterlogging and salinity are considered major factors influencing the distribution of salt marsh species (Adam 1990, Ungar 1998). Plants growing in waterlogged soils have to cope with lack of oxygen and accumulation of toxins (Kozlowski 1984). The salt-waterlogging stress is usually highest in the lower parts of the salt marsh (Chapman 1974, Kiehl 1997), whereas the upper parts experience decreased salt levels and longer periods of oxidation due to irregular inundation frequency and dilution by rain water (Armstrong et al. 1985, Perillo et al. 2009). However, in periods of high insolation and temperature, the soil solution in the upper marsh may become hypersaline due to increased evapotranspiration (De Leeuw *et al.* 1990). We found a decrease of nutrient content along the elevation gradients, with phosphorus, potassium and carbonate being highest in the lower marsh and which meets our expectations (Ranwell 1964, Adam 1990). However, the main environmental gradients (salt-waterlogging and nutrient gradient) were arranged almost orthogonally to each other, indicating that they were uncorrelated. The main driver of the nutrient gradient was the sand content of the soil which was higher on the island than on the mainland. Mainland soils of the Frisian coast usually contain more silt and are thus richer in nutrients, because they are shielded from the open sea water by barrier islands and mostly originate from land reclamation measures (Gray and Bunce 1972). Even the high marshes of the mainland were nutrient richer than the lower marshes of the island. Although many of the mainland marshes were utilized but not fertilized, the nutrient levels of these sites were also higher than the unused sites of corresponding elevation on the island.

In accordance with our first prediction, the three physiological traits conferring tolerance to salinity were most strongly correlated to the salt-waterlogging stress gradient. The more 'active' traits, dilution and exclusion of salt, increased with increasing groundwater salinity whereas the more 'passive' trait, morphological adaptations to osmotic stress, was prevalent at lower values of salinity. These opposite directions suggest that dilution and exclusion confer higher tolerance to salt than morphological responses.

Many authors have interpreted species distributions in salt marshes according to the physiological-ecological-amplitude concept by Scholten et al. (1987). Our analysis revealed four zones along the environmental gradients: (i) 'very benign' (low salt-waterlogging stress, high nutrient availability), (ii) 'benign' (low salt-waterlogging stress, low nutrient availability), (iii) 'stressful' (high salt-waterlogging stress, low nutrient availability) and (iv) 'very stressful' (high salt-waterlogging stress, low nutrient availability). We consider high salt-waterlogging stress in combination with high nutrient availability as more stressful than low salt-waterlogging stress in combination with low nutrient availability, as the influence of salinity and inundation frequency are considered the primary stress factors in salt marshes, from which salinity is regarded to as the 'master factor' (Rozema et al. 1985). Salinity affects biomass production and weight of the whole plant or roots and leaves of many salt marsh species and inundation reduces growth during submergence periods more in upper than in lower salt marsh plants (Groenendijk 1985, Rozema et al. 1985).

Following the physiological-ecological-amplitude concept we expected species of the 'very benign' and 'benign' habitats to show traits expressions conferring higher competitive effect, i.e. higher stem mass fraction, lower root mass fraction, higher SLA and lower LDMC and higher canopy height (Brouwer 1962b, Tilman 1988, Reich et al. 1991, Fonseca et al. 2000, Ackerly et al. 2002, Lavorel et al. 2007), but could only detect these trait expressions along the nutrient gradient, which included 'benign' and 'stressful' zones. It was evident that the physiological tolerance to salt stress decreased landward ('very benign', 'very stressful') but it was not evident that competitive ability increased landward. All traits commonly associated with strong competitive effect through shading either responded solely to the nutrient gradient (biomass allocation to stems, canopy height) or showed an opposite pattern to that expected on the salt-waterlogging stress gradient (SLA, LDMC and SDMC). Consequently, the question arises whether the undoubtedly strong competitive effect of e.g. *Elymus athericus* (Kuijper *et al.* 2005) is produced by the traits of the living plant. Based on a strong relationship between LDMC, C/N-ratio and decomposition in salt marshes of northwest Germany (Minden and Kleyer unpublished) we suggest that high LDMC, SDMC and C/N-ratio may lead to decreasing decomposition of dead biomass and thus to thick layers of litter in *Elymus athericus* stands, which might hamper the establishment of other species. We did not survey the influence of detritivores on litter decomposition, but we expect them to play a major role in litter accumulation and decomposition, respectively (see Buth & de Wolf 1985). Many studies reported an adverse effect of litter on biodiversity in various communities (see Xiong & Nilsson 1999).

For terrestrial plants, the leaf economics spectrum describes a trade-off from long life time of leaves, and low rates of photosynthesis and respiration to leaf traits with the potential of quick return of investments of nutrients, which proofed strongly correlated with SLA (Wright et al. 2004). The leaf economics spectrum was recently extended to a whole plant economics spectrum, including stem and leaf dry matter content, and related to a soil nutrient gradient (Freschet et al. 2010). Contrary to these findings and our expectations, results for SLA, LDMC and SDMC support the idea that leaf trait expressions of salt marsh species show an inverse relationship to those featured by other terrestrial plants. These traits only showed weak correlations to the nutrient gradient, but were strongly correlated to the salt-waterlogging stress gradient. Particularly LDMC increased with decreasing stress whereas SLA moderately decreased. This means that leaf traits like LDMC and SLA of salt marsh species are more constrained by salt and waterlogging than by nutrients and utilization. For instance, the species of the upper marsh experience physiological water deficits due to osmotic stress (*Elymus athericus* and *Festuca rubra*, Rozema et al. 1985) to which they respond by lignification of cell walls and increase in LDMC similar to plants experiencing frequent soil desiccation (Vendramini et al. 2002). In addition, high SDMC gives structural strength and supports upright position when turgor pressure is low. On the other hand, succulence and dilution of salt in the vacuole decreases LDMC and increases SLA by extending surface without increasing dry weight. The increased water content in leaves of some succulent lower marsh species contributed to higher SLA values, which was also found by Vendramini et al. (2002). However, we still expected a stronger relationship towards the nutrient gradient, which for example better explained SLA than climate data (e.g. mean annual precipitation) in a study by Ordoñez et al. (2009).

Considering partitioning of biomass on the nutrient gradient, our expectations were confirmed. As in other terrestrial and freshwater habitats, species growing at the nutrient rich salt marsh sites showed high values of stem mass fractions and canopy height (Gaudet and Keddy 1988, Westoby et al. 2002). These species shift their allocation to the stem to maintain dominance by occupying space and with that, compete for light (Kuijper et al. 2005). In contrast, when nutrient availability is low, the species showed high root mass fractions. Seasonal nutrient storage is essential for the success of species in more nutrient poor sites (Chapin 1980), because it enables growth when temperatures and radiation are most favorable and prevents species from day-to-day dependence upon available nutrients (Jeffrey 1964). Likewise, species showed high leaf mass fraction which can be attributed to a concurrent reduction in allocation to stems. Additionally, some of the species were rosette plants (e.g. *Plantago maritima*, *Triglochin maritimum*). Leaves of rosette plants grow near to ground-level and run a high risk of being shaded by taller plants. This feature may lead to impaired competitive ability, which restricts these species to nutrient poor sites (see also Van der Wal et al. 2000).

In general, our results are consistent with Brouwer's (1962b) theory and Tilman´s allocation model (1988) of a shift in allocation to plant organs that are responsible for capturing the resource that is limiting at a certain time, resulting in higher fractions of that organ relative to the whole plant body. Additionally, they support the meta-analysis of Poorter & Nagel (2000) showing that variations in biomass partitioning respond mainly to the nutrient gradient but less so to other environmental gradients, in our case to the salt-waterlogging stress gradient.

Conclusions

This study verified three common trait-environmental concepts for their applicability on salt marshes. The physiological-ecological-amplitude concept by Scholten et al. (1987) could only be supported for the nutrient gradiens but not for the salt-waterlogging stress gradient. Species growing in the 'stressful' zones of the lower marsh showed a higher competitive ability (canopy height, stem mass fraction) than the species of the upper marsh, although the latter were growing in the 'very benign' landward parts of the salt marsh. Leaf traits such as SLA and LDMC were more constrained by the influence of salt and waterlogging than by nutrient availability, which is opposite to the concept of the leaf economics spectrum (Wright et al. 2004) for other terrestrial plants. As in semi-arid environments (Vendramini et al. 2002), succulence led to an opposed response of leaf traits to environmental stress. Our results support Brouwer's (1962b) theory of functional equilibrium and Tilman's (1988) allocation model insofar as biomass fractions of belowground organs were high in nutrient poor sites, whereas species of nutrient rich sites showed high allocation to stem biomass to compete for light.

Acknowledgements

We thank the administration of the National park 'Niedersächsisches Wattenmeer' and 'Mellumrat e.V.' for their support during field work. Many thanks to G. Scheiffarth and M. Heckroth for supporting our work at Mellum and to J.P. Bakker for helpful comments on previous versions of the manuscript. This study was conducted as part of the TREIBSEL project and was supported by the 'II. Oldenburgischer Deichband' and the 'Wasserverbandstag e.V.' (NWS 10/05).

7 Environmental constraints on the C:N:P stoichiometry of plant organs in salt marshes

Vanessa Minden and Michael Kleyer,
in preparation

Abstract:

Vascular plants show a functional differentiation of their tissue, which is accompanied with differences in the elemental composition of the various organs. Furthermore, the availability of elements in the environment constrains nutrient uptake and assimilation rates. In our study we explore the C:N, C:P and N:P ratios of stems, leaves, diaspores and below-ground organs (roots and rhizomes) in salt marsh plants of Northwest Germany and evaluate the influence of environmental constraints on these element ratios.

We first conducted standardized major axis analysis (SMA) for a pooled dataset, which resulted in distinct patterns of isometric and anisometric SMA slopes. Subsequent bivariate-line fitting for the dataset split into four habitat types revealed that species of the frequently inundated lower marsh showed lower C:N ratios than those of the infrequently inundated upper marsh. We attributed this to elevated N-demand of the lower marsh species as an adaptation to salt stress. The influence of nutrient availability was detectable in increased C:P and N:P ratios in nutrient poor sites, indicating a stoichiometric response to P-availability. Across habitat types, leaves and diaspores showed higher elemental homeostasis than stems and below-ground organs. Any change in C:N ratios of below-ground organs and diaspores in response to the environment was accompanied by an even stronger change in stem C:N ratios, indicating a pivotal role of stems of herbaceous plants in ecosystem processes.

Key-words: bivariate line-fitting, ecological stoichiometry, plant strategies, salt marshes, scaling, standardized major axis analysis, tissue nutrient concentrations

7.1 Introduction

Photoautotrophs acquire elements (like C, N and P) necessary for growth and maintenance individually, mostly as inorganic matter, whereas heterotrophs mostly ingest elements bound in the biochemical structure of food (Ågren 2004). The latter may adjust their ingestion and the assortment of food in order to sustain elemental homeostasis (Frost et al. 2005). In contrast to heterotrophs, plants and algae show a wide variation in C:N:P ratios (Elser et al. 2000a, Hessen et al. 2004). This 'heterostasis' in element contents can be attributed as a response to environmental conditions (Vrede et al. 2004). Frost et al. (2005) showed examples for stoichiometric regulations of physiological processes in autotrophs, for example take-up of C in excess of metabolic demand or storage of elements in the vacuole. Yet metabolic demands for particular elements are affected by environmental supply and biological stressors (Frost et al. 2005), which has been shown by various studies for either the whole plant or certain plant organs, e.g. leaves (Wright et al. 2001, Güsewell and Koerselman 2002, Hessen et al. 2004, Ågren 2008).

The tissue-differentiation of higher plants into roots, stems, leaves and at a later stage diaspores has lead to a specialization of function, which is e.g. provision of water and nutrients by roots or synthesis of photosynthates by leaves. As a result, elemental composition differs between organs in multicellular organisms (Ågren 2008). For example leaves with high proportion of chloroplasts generally contain more N per unit dry mass than either stems or roots and show high N:P ratios (Epstein 1972). Flowering plants (shoots) may have lower N:P ratios than nonflowering plants of the same species and seeds generally have lower N:P ratios and higher N and P concentrations than vegetative structures (Fenner 1986, Güsewell 2004). An important factor influencing elemental composition within a plant organ is tissue age, as variation in nutrient requirements is related to ontogenetic changes (e.g. Sterner and Elser 2002). However, the elemental composition of tissue depends also on the availability of nutrients and water in the environment. Wright et al. (2001) found that species from drier sites had higher leaf N and P than those growing under more balanced water conditions and De Deyn et al. (2008) attributed increased biomass C:N content to a conservation-strategy when nutrients are limiting.

Against this background one can expect differences in C:N:P ratios of various plant organs relative to their function (i.e. structural support, leaf photosynthesis, provision of resources and regeneration). Kerkhoff et al. (2006) estimated scaling relationships for N and P concentrations and N:P ratios for different plant organs of woody and herbaceous plants and separated them into 'structural' (stems and roots) and 'metabolic' (leaves and reproductive) groups. However, they concentrated on N and P and neglected C, although these elements are strongly coupled (Knecht and Göransson 2004) and did not consider environmental constraints determining element distribution, but focused on phylogenetic and growth form association. Similarly, Craine et al. (2005) excluded environmental constraints when investigating N and P relationships (amongst others) of leaves and roots of grass species in four different regions.

Our study includes carbon, nitrogen and phosphorus as most important plant elements (Sterner and Elser 2002), factors all plant organs in considering their functional differentiation, relates the plant organ stoichiometry of a species pool to environmental conditions and focuses on a spatially well-defined area. Rather than scaling each element separately, we consider C:N, C:P and N:P ratios, because ratios were found to be less variable (at least N:P, see Güsewell and Koerselman 2002). We ask if plants shift the stoichiometry of their organs in response to the environment and if the relations between organs are kept constant. For instance, communities growing in resource-poor environments could show higher C:N:P values of all organs than those of resource-rich environments and this increase would be similar across all organs. Otherwise, plants could also shift stoichiometry between organs in different environments. In this case, for instance, the increase of the C:N:P ratio of stems across plants would be faster than that of leaves when the environment provides less resources.

The present study is located in salt marshes of Northwest Germany. Salt marshes consist of alluvial substrate ranging from clayey to sandy deposits. Lower parts of the salt marsh are flooded more regularly and for longer periods than upper marsh areas. The tidal rhythm of inundation washes in nutrients and organic compounds and influences the mineralization rate of plant litter (Adam 1990, Bakker et al. 1993). The major stress factors for plants colonizing salt marshes are salinity and inundation frequency, from which salinity is regarded to be the 'master factor' (Rozema et al. 1985). To prevent inactivation of enzymes and other essential structures under high salinities, salt marsh species synthesise nitrogen-requiring osmoprotectants (Steward et al. 1979, Tarczynski et al. 1993). These 'compatible osmotic solutes' often constitute a large part of the plant's nitrogen budget (Rozema et al. 1985) and we expect this to be reflected in the C:N:P ratios. To investigate the influence of the environment on the stoichiometry of salt marsh plants, we distinguished between four different habitats (1: nutrient rich, infrequently inundated, 2: nutrient rich, frequently inundated, 3: nutrient poor, infrequently inundated, 4: nutrient poor, frequently inundated). These habitats differed in regard to inundation frequency, level and salinity of groundwater and nutrient availability.

Specific expectations were as follows:

1. C:N:P ratios differ between plant organs pooled for all habitat types and show consistent patterns, i.e. 'structural' and 'metabolic' groups sensu Kerkhoff et al. (2006). N:P ratios should be lower in

diaspores than in vegetative structures (Güsewell 2004) and C:N ratios should be higher in stems and below-ground organs than in leaves (Epstein 1972, Sterner and Elser 2002).

2. C:N:P ratios of leaves, stems, diaspores and below-ground organs should differ between environments. For instance, C:N and C:P ratios of species from nutrient rich sites should be lower than those of species from nutrient poor sites (Wright et al. 2001, Lavorel et al. 2007) in leaves, stems and below-ground organs (see Freschet et al. 2010). Species growing in frequently inundated sites with high groundwater level and salinity should show lower C:N ratios and higher N:P ratios than species from infrequently inundated sites in leaves and stems (Bakker et al. 1993, van Wijnen and Bakker 1997). That is, plant organs should show consistent patterns of heterostasis in stoichiometry in response to the environment.

In this case, relationships between plant organs should either show a constant change in stoichiometry or shift in response to the environment. For instance, N:P ratios of leaves and stems could show a parallel increase when sites become more frequently inundated or, leaf N:P ratios could increase more strongly than those of stems. This would indicate different degrees in homoestasis across plant organs.

This study highlights the influence of environmental constraints on the elemental composition of all plant organs, which has not been done before in this detail. We were further able to demonstrate the role of stems in both the stoichiometry of plants and in ecosystem processes.

7.2 Materials and Methods

Study areas

Field work was carried out in three study areas along the coastline of Lower Saxony, Germany, and on the island of Mellum. The mainland areas were Leybucht (53°32'N, 7°07'E), Norderland (53°40'N, 7°19'E) and Jade Bight (53°26'N, 8°09'E), with 8, 89 and 32 plots, respectively. The prevailing soil substrate is clayey silt, loamy sand and loamy silt.

The island of Mellum (53°43'N, 8°08'E) originated from sand deposits transported by currents along the coast from The Netherlands and Northwest Germany (Pott 1995). It now belongs to the core zone of the Wadden Sea National Park of Lower Saxony. The predominant substrate is sand. On Mellum 33 plots were surveyed.

The mean annual temperature is 9°C and precipitation ranges from 770 – 830 mm/year (west to east, Deutscher Wetterdienst 2009).

Within the study areas, distribution of plots was based on random stratified sampling, with elevation as stratification criterion (see Krebs 1989). Random numbers were generated and provided distance measures from one corner of a plot to the corner of the next plot. All 152 plots were sampled during the growing season 2007.

Environmental parameters

To measure soil fertility soil samples were taken from each soil layer at each plot. Due to high groundwater levels, it was not possible to identify soil layers lower than 30 cm at many plots. Therefore, to make plots comparable all soils were sampled to a depth of 30 cm. From the soil samples the following parameters were investigated in the laboratory: bulk density (Schlichting et al. 1995), sand content (Ad-Hoc-AG Boden 2005), calcium carbonate ($CaCO_3$, according to Scheibler in Schlichting et al. 1995), plant available potassium (Flame photometer, Egnér et al. 1960) and phosphorus (Continuous Flow Analyser (CFA, Scalar Analytical, Breda, The Netherlands), Murphy and Riley 1962).

A drainage pipe (6.5 cm diameter, 162 in total) was buried 80 cm vertically in the ground at each plot. In these pipes the groundwater level was recorded biweekly from May to September 2007 at ebb tide, as well as the salinity content of the groundwater via conductivity ('WTWpH/Cond 340i/SET', WTW GmbH, Weilheim, Germany).

To record inundation frequencies and groundwater levels at high tide data loggers ("divers", ecoTech, Bonn, Germany) documented the salient water column in 18 drainage pipes in an hourly rhythm over the period of measurement. Additional four data logger recorded the pressure of the surrounding air, which was needed to calculate the relative pressure of water accumulating in the pipe. Inundation frequency was calculated from the elevation of all plots relative to the water level measured by the data loggers.

Plant species and nutrient stoichiometry

Almost all abiotic variables differed strongly between mainland and island salt marshes: phosphorous, potassium, calcium carbonate ($CaCO_3$), sand content of the soil, salinity of the groundwater and average groundwater level (Oneway ANOVA, p-value 0.05). Thus, mainland and island marshes were considered different habitats for this study. Consequently, we distinguished between taxa from the mainland and from the island (e.g. *Aster tripolium* from the mainland and *Aster tripolium* from the island). Plant material was collected for 14 mainland and 13 island taxa with approximately 9 individuals per species, resulting in 242 individuals in total. These individuals were selected from different plots covering the total range of species' occurrences within the environmental space.

Plants were collected at the peak of their generative stage, i.e. when seeds were ripened but not yet shed. Plants were dug out, roots and rhizomes were cleaned of soil material by rinsing off the soil substrate and roots of different individuals were carefully separated using tweezers. For chemical analysis, plants were separated into leaves, stems, diaspores and below-ground organs (i.e. roots and rhizomes). For each organ of each individual, plant material was grinded in a planetary mill for about 2 to 10 minutes at 300-400 revolutions ('pulverisette 7', Fritsch, Idar-Oberstein, Germany). For C:N analysis each sample was dried at 105 °C for 4-5 hours. 2-3 mg of material was filled into tin-tubes (0.1 mg precision balance CP 225 D, Sartorius, Goettigen, Germany) and analysed using CHNS-Analyser Flash EA (Thermo Electron Corporation, Oberhausen, Germany) following Allen (1989). Samples for plant phosphorus were dried at 70°C and 7-8 mg material were mixed with 0.2 ml nitric acid and heated for 6 hours at 95°C. After cooling down 0.03 ml hydrogen peroxide (32%) was added and the samples were heated again for 2 hours at 56°C. After repeated cooling the volume was completed to 1ml with water (bidest). Phosphorus content was measured using Continuous Flow Analyser (CFA) following Murphy and Riley (1962).

For some plant organs it was not possible to extract enough material for nutrient and carbon measurement (e.g. diaspores of *Salicornia europaea*); data from those cases were left out. At each plot, species composition and abundance was recorded with frequency analysis using a 1 x 1 m frame subdivided into 100 grids of 0.1 x 0.1 m.

Statistical analysis

Groundwater level

At all plots groundwater level was recorded biweekly during low tide. A regression was conducted with paired values of the hourly data produced by the 18 loggers and the biweekly data of the groundwater levels of all plots. Mean values of both groundwater and logger data over the measurement period were generated and used for linear regression analysis. The regression function was used to adjust values of mean groundwater level of all other plots to include information about high tide.

Bivariate line-fitting methods

We used standardized major axis (SMA) slope-fitting techniques to explore bivariate relationships between the plant traits, as both the X and Y variables had variation with them due to measurement error, which makes common linear regression inappropriate (Sokal and Rohlf 1994, Warton et al. 2006). The aim of line-fitting technique is to summarize the relationship between two variables by minimizing the residuals in both variables (Kerkhoff et al. 2006), rather than predict Y from X, for which ordinary least square (OLS) regression would be the adequate technique (Niklas 2006).

When dividing the dataset according to different environments, the clouds of points in an SMA diagram may show different alignments. First, two sites may show a shift along the common slope (site 1 and site 2, shift 1, Fig. 7-1). If so, then both sites contain species sharing the same ratios between two traits, however species from one site generally have higher trait expressions (i.e. higher values on the X-axis with subsequently higher Y-values). Species from two sites showing difference in one trait at a given second trait-value (shift in Y combined with no change in X) show a shift in elevation amongst two lines (site 1 and site 3, shift 2) and will show different ratios between the traits.

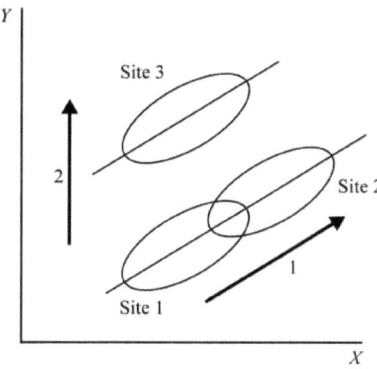

Figure 7-1: Possible alignments of clouds of points sharing a common slope for traits X and Y, adopted from Wright et al. (2001). Site 1 and site 2 show a shift along a common slope (shift 1), whereas site 3 shows a higher elevation than site 1 (shift 2).

We conducted two separate SMA analyses (Fig. 7-2). First, to test whether C:N, C:P and N:P ratios differ between plant organs, we analysed each possible trait combination regardless of habitat type (Fig. 7-2, dark box). We then separated our data set into plants groups occurring in four different habitat types. We repeated SMA analysis for each trait combination of each habitat type to find shifts along SMA slopes and in elevations between habitat types (Fig. 7-2, light grey box).

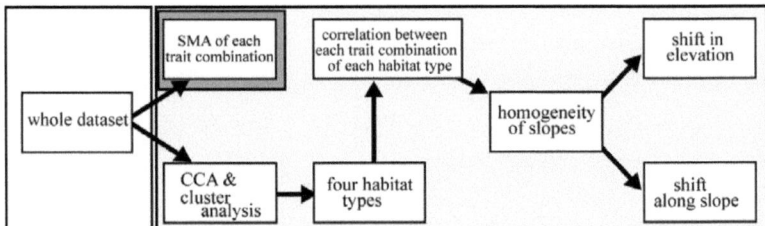

Figure 7-2: Flowchart of statistical analysis. First, SMA was conducted for the whole dataset and for each trait combination of each habitat type (dark box). Second, bivariate line fitting-techniques were carried out for each trait combination for each habitat type (light grey box).

We then tested for significant correlations between two traits of each group at $p < 0.05$ (Pearson's r). The further analysis consisted of three steps and only considered the relationships of the pairwise C:N, C:P and N:P ratios of the plant organs where significant correlation existed. First, we tested for homogeneity of slopes with $p > 0.05$. Where a common slope could be fitted, we tested for differences in elevation and shift along the axis (steps two and three; Wald statistic). All data were log-transformed for analysis and analyses were performed using 'smatr-package' (Warton) for the computer software R (The R Foundation for Statistical Computing 2008).

7.3 Results

CCA of environmental parameters, species clustering and characteristics of habitat types

CCA-analysis explained 24% of the inertia, of which 19% were explained by the first two axes (Fig. 7-3). The first set of variables, i.e. potassium, phosphorus, $CaCO_3$ and sand spanned a 'nutrient' gradient from nutrient rich to nutrient poor sites and also separated mainland from island plots. A second 'salt-waterlogging stress' gradient, uncorrelated to the first gradient, was determined by inundation frequency, groundwater salinity and groundwater level. It separated highly inundated sites with high levels of salty groundwater from infrequently inundated sites with low groundwater levels.

Four clusters were positioned in the interspace between the major ordination gradients, i.e. cluster 1 contained species from infrequently inundated mainland sites with high nutrient content of the soil (Fig. 7-3, Table 7-1), cluster 2 was positioned in nutrient rich mainland sites with high influence of salty groundwater and inundation. Species of cluster 3 and 4 grew in nutrient poor soils of the island and were positioned at the opposite sides of the water gradient (cluster 3 in infrequently, cluster 4 in frequently inundated sites). The resulting habitats were: nutrient rich, infrequently inundated sites (cluster 1) vs. nutrient rich, frequently inundated sites (cluster 2, both mainland) vs. nutrient poor, infrequently inundated sites (cluster 3) vs. nutrient poor, frequently inundated sites (cluster 4, both island).

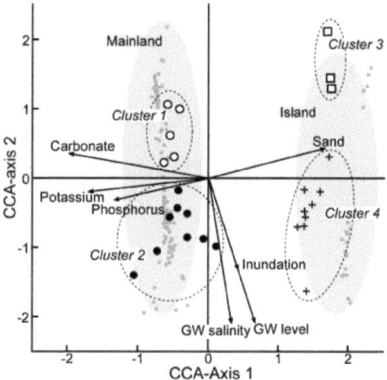

Figure 7-3: Ordination diagram of the first two axes of CCA-analysis displaying the environmental variables and position of species of four clusters as results of cluster analysis. Small grey dots represent plots, GW: groundwater.

Table 7-1: Description of four habitat types (1-4) with mean values for each environmental parameter with standard deviations in brackets

Parameter	Unit	(1) Nutrient rich, infrequently inundated sites	(2) Nutrient rich, frequently inundated sites	(3) Nutrient poor, infrequently inundated sites	(4) Nutrient poor, frequently inundated sites
Phosphorus	kg/ha	199.45 (55.91)	190.29 (54.69)	65.54 (18.29)	125.48 (61.81)
Potassium	kg/ha	1,796.62 (594.10)	1,979.41 (489.69)	696.35 (138.49)	939.25 (384.27)
CaCO$_3$	t/ha	210.43 (55.71)	174.86 (65.01)	14.49 (16.78)	17.02 (27.59)
Sand content	%	46.70 (20.66)	43.59 (20.52)	79.41 (8.97)	65.11 (19.48)
Groundw. level	cm	-38.15 (11.89)	-23.09 (10.14)	-36.17 (13.39)	-17.44 (12.18)
Groundw. salinity	PSU	21.33 (3.60)	25.08 (3.43)	21.42 (3.36)	26.40 (1.52)
Inundation	hours	38.37 (80.17)	214.05 (293.60)	169.21 (178.51)	262.26 (377.55)

SMA of C:N, C:P and N:P ratios of each organ combination for the pooled dataset

Considering C:N:P ratios for the whole dataset, C:N ratio was highest in stem and below-ground mass, however, standard deviation was also highest there, compared to leaf and diaspore C:N content (Table 7-2, upper part). The latter two showed lower C:P ratios than both stem and below-ground mass. N:P ratios of the four plant organs were more balanced, however, below-ground organs showed the lowest values. Across all species, the standard deviations of leaves and diaspores were lower than of stems and below-ground organs.

SMA regression of the whole dataset was conducted to verify whether the C:N:P ratios differ between the plant organs. Four relationships were not significant (Table 7-2, lower part). We found isometric relationships (slope ~1) between five organ combinations, i.e. the ratio of one organ increased linearly with another organ. Those combinations were the C:N ratios of stem and below-ground and leaf and diaspore, C:P ratios of leaf and diaspore and N:P ratios of stem and below-ground and leaf and diaspore mass. All other relationships were anisometric (positive slope ≠ 1), i.e. ratios of one organ increased either faster or more slowly with another organ. The patterns of isometry and anisometry of the slopes of these relationships follow those of Kerkhoff et al. (2006, expectation 1).

Table 7-2: Upper part: Mean and standard deviation of C:N, C:P and N:P ratios for each plant organ (no log-transformation). Lower part: Statistics of SMA slopes (log-log) of each pairwise combination between plant organs of the whole dataset. Slope of the regression, 95% confidence intervals in parentheses, r^2 and P-value. Isometric slopes are shown in bold. NS indicates a non-significant relationship.

Mean, standard deviation		C:N	C:P	N:P
stem		60.35, 29.36	551.82, 358.85	9.91, 5.69
leaf		23.82, 7.69	245.01, 115.16	10.47, 3.51
diaspore		20.57, 5.77	189.88, 85.08	9.35, 3.63
below-ground		54.98, 28.05	352.54, 158.66	7.15, 2.65
X	Y	C:N	C:P	N:P
stem	leaf	0.69 (0.60, 0.79), 0.23, <0.001	0.73 (0.62, 0.84), 0.19, <0.001	0.69 (0.57, 0.83), 0.17, <0.001
stem	diaspore	0.57 (0.49, 0.66), 0.28, <0.001	0.78 (0.69, 0.89), 0.33, <0.001	0.69 (0.61, 0.78), 0.33, <0.001
stem	below-ground	**1.09 (0.93, 1.30), 0.18, <0.001**	NS	**0.90 (0.68, 1.18), 0.11, <0.001**
leaf	diaspore	**0.88 (0.73, 1.06), 0.31, <0.001**	**1.08 (0.91, 1.32), 0.26, <0.001**	**1.03 (0.87, 1.26), 0.17, <0.001**
leaf	below-ground	1.52 (1.29, 1.78), 0.15, <0.001	1.29 (0.93 1.73), 0.05, 0.03	NS
diaspore	below-ground	1.60 (1.35, 1.91), 0.11, <0.001	NS	NS

Comparison of C:N:P scaling relationships between organs and different environments

The analysis was repeated for each trait combination for each habitat type (Appendix 1). Mean values of C:N, C:P and N:P ratios of plant organs were similar to those pooled for all habitat types, i.e. C:N and C:P content for stem and below-ground were higher than for leaf and diaspore mass, whereas N:P ratios were more balanced but always lowest in diaspore mass. Across all organs, C:N ratios were highest in the infrequently inundated habitats (Fig. 7-4, clusters 1 and 3), although not always with significant differences to the frequently inundated habitats. C:P ratios did not vary that strongly between habitat types and N:P ratios showed similar patterns to those of C:N, at least for stem and leaf mass.

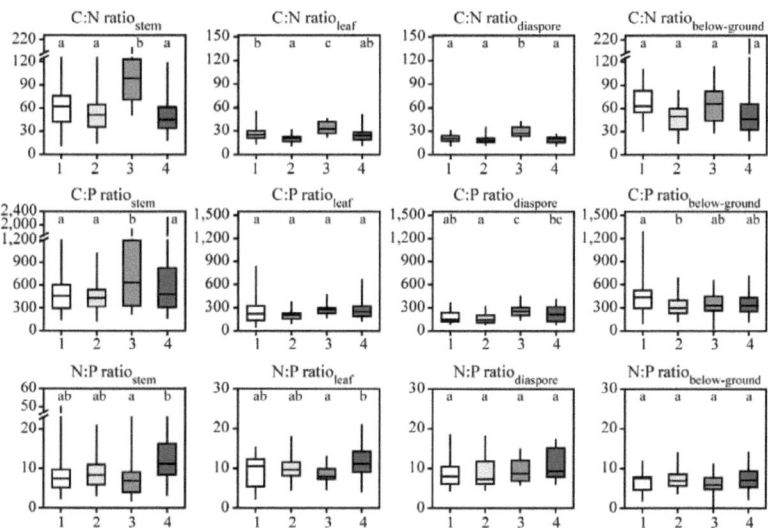

Figure 7-4: Boxplots of C:N, C:P and N:P ratios of each plant organ for each habitat type. For abbreviations of habitat types see Table 7-1. Significant differences (ANOVA, P-value < 0.05) between habitat types are indicated by different letters in each box.

Significant correlations of C:N, C:P and N:P ratios of each plant organ (P-value < 0.05, Appendix 7-1) were examined in 37 of 72 possible cases. Out of those, 27 pairwise relationships could be analysed further (Appendix 7-2). A common slope could be fitted in eighteen cases, i.e. the null hypotheses of a common slope was not rejected. For those combinations we tested for equal elevation and for shift along the common axis.

Considering scaling relationships, seven slopes showed isometry and 30 anisometry (Appendix 7-1). Consistent patterns comparable to those of the pooled dataset could not be detected.

A shift along the common slope was found for C:N ratios between stem and leaf mass (cluster 1 vs 2). Species of nutrient rich, frequently inundated sites showed lower C:N ratios than species of nutrient rich, infrequently inundates sites (Fig. 7-5a). The same result was obtained for the C:N ratios between stem and diaspore mass (Fig. 7-5b).

C:N ratios of stem and below-ground mass shifted in both elevation and along the slope between cluster 2 and 3, and 3 and 4. Species growing in nutrient poor, infrequently inundated sites showed a higher C:N ratio and allocated less N and more C to their stems than to their below-ground organs, as compared to species of nutrient rich, frequently inundated sites (Fig. 7-4c). The same counts for species from nutrient poor, infrequently inundated sites compared to species from nutrient poor, frequently inundated sites. Also, species from nutrient poor, frequently inundated sites showed lower C:N ratios of stem and leaves than species from nutrient poor, infrequently inundated sites (cluster 3 vs. 4, Fig. 7-4 d).

Considering C:P ratios, analysis found shifts along the slope for stem and leaf mass in cluster 1 vs 4, and 2 vs 4 (Fig. 7-4e). In both cases, species from cluster 4 (nutrient poor, frequently inundated sites) showed higher C:P values than those from nutrient rich, infrequently inundated (cluster 1) and nutrient rich, infrequently inundated sites (cluster 2). The same pattern could be detected for C:P ratios of stem and diaspore mass (Fig. 7-4f). Species growing in nutrient poor sites had higher C:P ratios than those from nutrient rich sites, both in infrequently inundates sites (cluster 1 vs 3) and frequently inundates sites (cluster 2 vs 4).

Under the same flooding frequency and differing in nutrient availability, species growing in nutrient poor soils showed higher N:P ratios in stem and leaf mass (Fig. 4g) and lower N:P ratio in below-ground organs for a given stem mass (Fig 4i).

Analysis obtained two significant results for N:P ratios for stem and diaspore mass, which were both between stem and diaspore mass (Fig. 4h). First, cluster 3 and 4 showed a shift along the slope and a shift in elevation, meaning that species of nutrient poor, frequently inundated sites showed higher N:P ratios than species of nutrient poor sites exposed to lesser flooding frequency. A shift along the slope was detected between cluster 2 and 4. Species from nutrient rich, frequently inundated sites showed lower N:P ratios than did species from nutrient poor, frequently inundated sites.

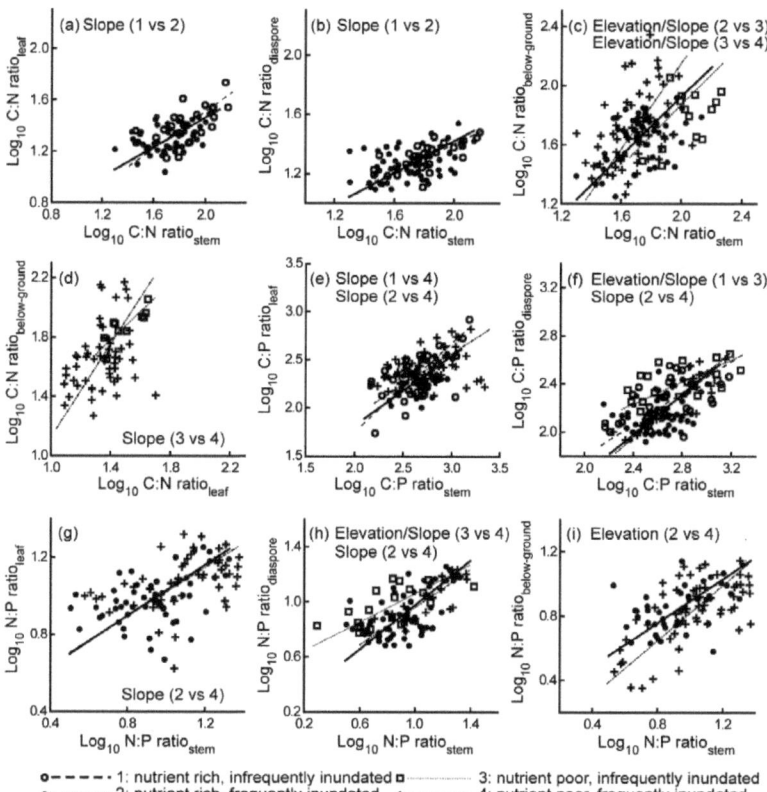

Figure 7-5: Standardized major axis regression relationships between C:N and N:P ratios of stem, leaf, diaspore and below-ground biomass for four habitat types. Habitat types are nutrient rich, infrequently inundated sites (cluster 1: open circle, dashed line), nutrient rich, frequently inundated sites (cluster 2: closed circle, solid line), nutrient poor, infrequently inundated sites (cluster 3: open square, dotted line) and nutrient poor, frequently inundated sites (cluster 4: cross, dash-dot line). Slope: shift along slope, Elevation: shift in elevation.

7.4 Discussion

C:N:P relationships between tissue types

We expected C:N:P ratios of the different plant organs to be related to their functions (photosynthesis, supporting structure, storage, reproduction, see Wright et al. 2001, Kerkhoff et al. 2006). For example leaves should show lower C:N and C:P ratios and higher N:P ratios than stems. This is because photosynthesis requires large amounts of proteins (e.g. Rubisco), which are N-based, and proteins presuppose rRNA, which is P-based (Ågren 2004). Tissue with high allocation to chloroplasts compared to other structures have high N:P ratio (Epstein 1972, Sterner and Elser 2002). On the other hand, supporting structures such as stems predominantly consist of cellulose and lignin, which are C-rich and nutrient poor (McGroddy et al. 2004). Our analyses showed that C:N and C:P ratios of stem mass is higher than of leaves and N:P ratio was lower, both for the pooled dataset and for each habitat type and SMA analysis revealed allometric relationships between these plant organs (Table 7-2, Appendix 1), which clearly confirms this expectation.

Güsewell (2004) states that seeds have lower N:P ratios than vegetative structures. However the effects of nutrient supply on nutrient concentration in seeds may vary between plant species, from either constant concentrations (Fenner 1986) to lower concentration (lower P concentrations in P limited sites, Lewis and Koide 1990). Our results showed that diaspores indeed have a lower C:N ratio than leaves, and N:P ratios were equal and lower in diaspores than in stems, respectively.

There were distinct patterns of isometric and anisometric relationships between the C:N:P ratios of the plants organs when exploring the pooled dataset. We found isometric scaling when C:N:P ratios of stem and below-ground as well as leaf and diaspores were compared. However, anisometric scaling was observed at all other significant combinations, like stem vs. leaf, stem vs. diaspores, leaf vs. below-ground and so forth. These results support the findings of Kerkhoff *et al.* (2006) who used a much larger dataset spanning species from arctic to tropical regions. They applied the term 'structural' to group stems and below-ground organs and 'metabolic' to group leaves and diaspores and found that scaling within these groups tended to yield isometric relationships whereas scaling between these groups was anisometric. Finding similar patterns in our case study supports the conclusion of Kerkhoff *et al.* (2006) that these patterns advert to general constraints or allocation rules, which may govern the partitioning of nutrients among plant organs. However, these patterns could not be clearly reproduced in the four smaller subsets of our data. Reason for this may be increased influence of residuals and outliers, which may have perished in the pooled dataset.

Within the pooled dataset the scaling relationships between stem and leaf, and stem and diaspores were < 1, which means that C:N, C:P and N:P ratios are always higher in stems than in leaves and diaspores and scale about the $^{3/4}$–power for C:P and N:P and about $^{2/3}$-power for C:N. C:N, C:P and N:P ratios were higher in below-ground organs than in leaf and diaspores and slopes were > 1. Niklas and Enquist (2002) found scaling relationships between leaves and stems, and leaves and roots of 1 when observing biomass relationships. It seems that stoichiometric relationships are slightly different from biomass relationships, however, there is need for more data to substantiate these results.

The influence of environmental constraints on C:N:P ratios

The physiological mechanisms of salt marsh plants are set apart from other terrestrial plants insofar, that part of their nitrogen and carbon budget is used for the synthesis of 'compatible osmotic solutes' (Rozema *et al.* 1985). As this adaptation to salt stress may influence their C:N:P stoichiometry, a short review of the underlying mechanisms seems appropriate at this point. The osmotic potential of the external medium in which salt marsh plants grow is by times very low, due to high amounts of solute ions. Osmotic adjustment to the external concentration would result in toxic levels. Thus, Na^+ and Cl^- are compartmentalized, predominantly in vacuoles so that concentrations in the cytoplasm are kept within tolerable limits (Gutknecht and Dainty 1969, Wyn Jones and Gorham 2002). To prevent dehydration of the cytoplasm and inactivation of enzymes and other essential structures, plants accumulate metabolically 'compatible' solutes (Borowitzka 1981). A range of molecules such as sugars, sugar alcohols, amino acids and betaines have been identified to fulfil this function in halophytes, which are associated with elevated C and N costs (Hasegawa et al. 2000, Rhodes et al. 2002, Flowers and Colmer 2008). This demand of, in particular N, should be reflected in the C:N:P stoichiometry of salt marsh plants.

Examination of the C:N, C:P and N:P values of the split dataset (Appendix 1) reveals that standard deviation of leaves and diaspores is, for the most part smaller than for stem and below-ground mass. This might signal a more homeostatic regulation in the 'metabolic' than in the 'structural' organs.

Mean level and salinity of the groundwater as well as inundation frequency were highest in the frequently inundated, lower marsh (Table 7-1), hence we expected C:N ratios of species growing in those parts to be lower than that of species colonizing the infrequently inundated, higher parts of the marsh. Indeed, our expectations were met in four cases (Figure 5 a, b,c), in which C:N content of the tissue from species in the frequently inundated habitats were lower than that of species growing in the infrequently inundated habitats. Reason for this might be the increased demand of nitrogen of species of the lower marsh to produce compatible osmotic solutes in sufficient quantity. Another reason might be the increased lignin content of tissue of upper marsh species as an adaptation to salt stress, which also resulted in lower decomposition rates (Minden and Kleyer unpublished work).

In habitats with lower nutrient availability, species show higher C:P and N:P ratios (Fig. 5e, f, g, h) compared to those from nutrient richer sites. It seems that the reason for this is the decreased amount of phosphorus in the nutrient poor sites, which in reverse increases C:P and N:P ratios. Whereas C:N ratio of salt marsh plant seems more constrained by the influence of water (inundation) stress, C:P and N:P ratios respond more strongly to differences in nutrient availability. Considering N:P ratios, we found higher ratios in species from the lower marsh compared to the higher marsh, when comparing stem and diaspore mass (Figure 5h), which may also be due to increased production of osmoprotectants of species of the lower marsh.

Interestingly, we only detected in shifts in elevation when stems are compared to other organs. From the nutrient rich, frequently inundated sites (lower mainland marshes) to the nutrient poor, infrequently inundated sites (upper island marshes), N decreased more strongly in stems than in below-ground organs and in diaspores, either as a function of C or P. All other relationships between organs did not shift in elevation in response to the environment. This suggests that the stoichiometry of stems is most responsive to environmental changes in salt marshes whereas other organs achieve higher homoestasis across environments to perform their functions (see also Kerkhoff et al. 2006). It is possible that stems demand lower amounts of osmoprotectants on upper marshes and invest more carbon in structural tissues as compared to the more 'metabolic' organs, whereas on lower marshes high nitrogen recycling in the phloem is necessary to allow the synthesis of larger amounts of osmoprotectants.

Conclusions

The results of this study clearly demonstrate the existence of distinct patterns of C:N:P ratios of plant organs in relation to their functions. We could detect patterns of isometric and anisometric relationships between 'structural' (stem, below ground) and 'metabolic' organs (leaf, diaspore) when investigating the pooled dataset. This suggests the existence of a set of common rules across many plants that govern the partitioning of nutrients among plant organs (Kerkhoff et al. 2006). Whether these rules depend on environmental conditions or on basic physiological requirements have rarely been investigated. Leaves and diaspores showed a higher degree of homoestasis in C:N:P ratios indicating that functioning of these organs is more constrained by basic physiological requirements than that of stems and below-ground organs. However, we could also detect an increased nitrogen demand of the lower marsh species, presumably due to enhanced production of N-demanding compatible osmotic solutes in order to avoid inactivation of enzymes and dehydration of the cytoplasm. In most bivariate relationships, both organs displayed the same stoichiometric responses to the environment. Stems only showed a stronger response than other organs. Thus, changes in plant community composition towards plants with increased C:N ratios in e.g. belowground organs would likely entail an even greater increase in the C:N ratio of stems. This indicates the significance of the elemental composition of plant stems on environmental gradients that might also have profound effects on ecosystem functions such as decomposition or biogeochemical cycles.

Acknowledgments

We thank the administration of the National park 'Niedersächsisches Wattenmeer' and 'Mellumrat e.V.' for their support during field work. Many thanks to H. Timmermann and J. Spalke for contributing their data, and to G. Schweiffarth and M. Heckroth for supporting our work at Mellum.

We also thank Helmut Hillebrand for comments to an earlier version of the manuscript. This study was conducted as part of the TREIBSEL project and was supported by the 'II. Oldenburgischer Deichband' and the 'Wasserverbandstag e.V.' (NWS 10/05).

Appendix 7-1: Upper part: Mean and standard deviation of C:N, C:P, and N:P ratios for each plant organ (no log-transformation) within each habitat type (i.e. each cluster). Lower part: Statistics of SMA slopes (log-log) of each pairwise combination between plant organs for each habitat type. Correlation coefficients, slope of the regression, 95% confidence intervals in parentheses, r^2 and P-value. Isometric slopes are shown in bold. NS indicates a non-significant relationship.

		Nutrient rich, infreq. inundated (1)			Nutrient rich, freq. inundated (2)		
		C:N	C:P	N:P	C:N	C:P	N:P
stem		67.5, 33.7	506.5, 325.1	8.8, 7.9	55.1, 22.4	447.8, 175.6	8.9, 4.1
leaf		26., 8.8	254.3, 158.7	9.6, 3.7	20.4, 4.5	201.9, 61.7	10.1, 3.0
diaspore		20.7, 4.7	178.4, 75.1	8.7, 3.3	18.6, 4.4	160.6, 60.5	8.9, 3.8
below-ground		67.9, 22.7	461.2, 297.2	6.7, 2.8	46.6, 16.1	321.8, 118.6	7.4, 2.3
C:N		Nutrient rich, infreq. inundated (1)			Nutrient rich, freq. inundated (2)		
stem	leaf	0.64, 0.72 (0.48, 0.97), 0.40, <0.01			0.36, 0.57 (0.45, 0.73), 0.13, <0.01		
stem	diasp.	0.69, 0.47 (0.36, 0.63), 0.47, <0.01			0.36, 0.53 (0.43, 0.64), 0.13, <0.01		
stem	bel-grd.	NS			0.49, **1.00 (0.81, 1.26), 0.24, <0.01**		
leaf	diasp.	0.62, 0.66 (0.47, 0.88), 0.38, <0.01			0.47, **1.00 (0.75, 1.29), 0.22, <0.01**		
leaf	bel-grd.	NS			NS		
diasp.	bel-grd.	NS			NS		
C:P		Nutrient rich, infreq. inundated (1)			Nutrient rich, freq. inundated (2)		
stem	leaf	0.6, **0.92 (0.68, 1.18), 0.36, <0.01**			0.33, 0.73 (0.55, 0.91), 0.11, 0.01		
stem	diasp.	0.52, 0.65 (0.49, 0.90), 0.27, <0.01			0.36, 0.83 (0.66, 1.03), 0.13, <0.01		
stem	bel-grd.	NS			NS		
leaf	diasp.	0.61, 0.69 (0.53, 0.97), 0.37, <0.01			0.39, **1.19 (0.96, 1.51), 0.16, <0.01**		
leaf	bel-grd.	NS			NS		
diasp.	bel-grd.	NS			-0.43, -0.98 (-1.23, -0.76), 0.18, <0.01		
N:P		Nutrient rich, infreq. inundated (1)			Nutrient rich, freq. inundated (2)		
stem	leaf	NS			0.42, 0.65 (0.51, 0.82), 0.18, <0.01		
stem	diasp.	NS			0.59, 0.81 (0.68, 0.95), 0.34, <0.01		
stem	bel-grd.	NS			0.47, 0.68 (0.54, 0.84), 0.22, <0.01		
leaf	diasp.	0.42, 0.72 (0.48, 1.03), 0.18, 0.03			0.35, 1.32 (1.05, 1.73), 0.13, 0.01		
leaf	bel-grd.	NS			NS		
diasp.	bel-grd.	NS			0.43, 0.75 (0.60, 0.91), 0.12, <0.01		
		Nutrient poor, infreq. inundated (3)			Nutrient poor, freq. inundated (4)		
		C:N	C:P	N:P	C:N	C:P	N:P
stem		99.1, 38.8	740.7, 479.9	7.4, 5.4	51.2, 19.6	627.4, 436.7	12.1, 5.5
leaf		33.3, 8.3	277.2, 89.4	8.5, 2.3	24.2, 7.6	274.6, 123.4	11.6, 3.7
diaspore		28.5, 6.2	269.3, 84.7	9.7, 2.9	18.8, 4.1	218.4, 107.7	11.1, 3.6
below-ground		63.4, 24.6	347.1, 153.2	5.9, 2.5	56.8, 36.9	356.9, 146.0	7.3, 2.9
C:N		Nutrient poor, infreq. inundated (3)			Nutrient poor, freq. inundated (4)		
stem	leaf	NS			NS		
stem	diasp.	NS			NS		
stem	bel-grd.	0.54, **1.03 (0.69, 1.50), 0.29, 0.03**			0.4, 1.46 (1.13, 1.90), 0.16, <0.01		
leaf	diasp.	NS			NS		
leaf	bel-grd.	0.81, **0.95 (0.49, 1.35), 0.65, <0.01**			0.32, 1.56 (1.12, 1.99), 0.09, 0.02		
diasp.	bel-grd.	NS			0.46, 1.84 (1.16, 2.62), 0.21, 0.03		
C:P		Nutrient poor, infreq. inundated (3)			Nutrient poor, freq. inundated (4)		
stem	leaf	NS			0.33, 0.64 (0.51, 0.79), 0.11, 0.01		
stem	diasp.	0.52, 0.46 (0.32, 0.59), 0.27, 0.02			0.84, **0.89 (0.74, 1.10), 0.71, <0.01**		
stem	bel-grd.	NS			NS		
leaf	diasp.	NS			0.67, 1.90 (1.46, 2.81), 0.45, <0.01		
leaf	bel-grd.	NS			0.51, 1.21 (0.97, 1.45), 0.26, <0.01		
diasp.	bel-grd.	NS			0.62, 0.82 (0.57, 1.10), 0.35, <0.01		
N:P		Nutrient poor, infreq. inundated (3)			Nutrient poor, freq. inundated (4)		
stem	leaf	NS			0.43, 0.63 (0.45, 0.87), 0.18, <0.01		
stem	diasp.	0.64, 0.52 (0.38, 0.78), 0.41, <0.01			0.83, 0.68 (0.56, 0.89), 0.69, <0.01		
stem	bel-grd.	NS			0.64, 0.86 (0.71, 1.03), 0.41, <0.01		
leaf	diasp.	NS			0.73, 1.50 (1.17, 1.94), 0.53, <0.01		
leaf	bel-grd.	NS			0.29, -1.28 (0.92, 1.74), 0.08, 0.03		
diasp.	bel-grd.	NS			NS		

Appendix 7-2: Results of standardized major regression analysis for all significantly correlated trait combinations for four habitat types. Significant shifts in elevation and along the slope are shown in bold. NA indicates that there was insufficient slope homogeneity between the traits to justify further analysis. Habitat types: 1 = nutrient rich, infrequently inundated; 2 = nutrient rich, frequently inundated; 3 = nutrient poor, infrequently inundated; 4 = nutrient poor, frequently inundated. Homog.= homogeneity

Trait pair	Habitat type	Slope homog. (p)	Shift in elevation (p)	Shift along slope (p)
C:N				
stem vs leaf	1 vs 2	0.27	0.52	**<0.01**
stem vs diaspore	1 vs 2	0.51	0.89	**<0.01**
stem vs below-ground	2 vs 3	0.89	**0.03**	**<0.01**
stem vs below-ground	2 vs 4	0.03	NA	NA
stem vs below-ground	3 vs 4	0.19	**<0.01**	**<0.01**
leaf vs diaspore	1 vs 2	0.04	NA	NA
leaf vs below-ground	3 vs 4	0.06	0.91	**<0.01**
C:P				
stem vs leaf	1 vs 2	0.25	0.20	0.59
stem vs leaf	1 vs 4	0.09	0.71	**0.04**
stem vs leaf	2 vs 4	0.54	0.19	**<0.01**
stem vs diaspore	1 vs 2	0.25	0.27	0.25
stem vs diaspore	1 vs 3	0.16	**0.03**	**<0.01**
stem vs diaspore	1 vs 4	0.12	0.39	0.12
stem vs diaspore	2 vs 3	0.01	NA	NA
stem vs diaspore	2 vs 4	0.65	0.91	**0.01**
leaf vs diaspore	1 vs 2	0.01	NA	NA
leaf vs diaspore	1 vs 4	0.01	NA	NA
leaf vs diaspore	2 vs 4	0.03	NA	NA
diaspore vs below-ground	2 vs 4	0.43	0.09	0.53
N:P				
stem vs leaf	2 vs 4	0.85	0.89	**<0.01**
stem vs diaspore	2 vs 3	0.04	NA	NA
stem vs diaspore	2 vs 4	0.29	0.53	**<0.01**
stem vs diaspore	3 vs 4	0.22	**<0.01**	**<0.01**
stem vs below-ground	2 vs 4	0.18	**<0.01**	0.3
leaf vs diaspore	1 vs 2	<0.01	NA	NA
leaf vs diaspore	1 vs 4	<0.01	NA	NA
leaf vs diaspore	2 vs 4	0.54	0.35	0.3

8 Testing the effect-response framework: key response and effect traits determining aboveground biomass of salt marshes

Vanessa Minden, Michael Kleyer
submitted to Journal of Vegetation Science

Abstract

Question: How do species' traits respond to environmental conditions and what is their effect on ecosystem properties?

Location: Salt marshes, Northwest Germany

Methods: On 113 plots along the mainland coast and on one island we measured environmental parameters (soil nutrient content, inundation frequency, groundwater level and salinity), collected traits from 242 individuals (specific leaf area (SLA), whole plant C:N ratio, and dry weights of plant organs) and sampled aboveground biomass as an ecosystem property. We constructed a path model combining environmental parameters, functional traits at community level and aboveground biomass, which was tested against a dependence model using path analysis; model fitness was evaluated by structural equation modelling (SEM).

Results: The final model showed good consistency with the data and points out the major role of groundwater level and salinity and nutrient availability as most important factors influencing biomass allocation in salt marshes. Aboveground living biomass was mostly determined by stem biomass, which was mediated through an allometric allocation of biomass to all other plant organs, including leaf mass. C:N-ratio and SLA were the major driver for dead biomass.

Conclusions: We emphasise an indirect link between standing biomass and environmental conditions and recognize stem biomass, plant C:N-ratio an SLA as keystone markers of species' functioning in determining the relationship between environment and ecosystem properties.

Keywords: allometric relationships, trade-off, effect trait, response trait, path analysis, structural equation modelling, salt marshes

8.1 Introduction

In recent years there has been growing consensus that the use of functional traits of plants, rather than taxonomic specification, can assist in explaining vegetation-related ecosystem properties and functions (Hooper et al. 2005). Lavorel and Garnier (2002) developed a conceptual framework that focuses on the prediction of changes in ecosystem processes using plant functional traits. It distinguishes between functional response traits (set of traits associated with the response of plants to environmental factors, such as allocation of biomass as a response to nutrient availability, Gitay and Noble 1997, Walker et al. 1999) and functional effect traits (set of traits related to the effect of plants on the environment, such as the effect of plant traits on the decomposability of litter, Díaz and Cabido 2001). However, response and effect traits sometimes overlap, i.e. traits can both explain plant responses to environmental conditions and effects on the ecosystem (Lavorel and Garnier 2002). This is said to be particularly the case in the use of resources (Díaz and Cabido 2001) and primary productivity (Lavorel and Garnier 2002). Therefore, Suding et al. (2008) proposed to simultaneously identify traits that respond to environmental gradients and traits affecting ecosystem properties. In our paper we apply this approach to salt marshes of Northwest Germany by linking plant functional traits with abiotic conditions and ecosystem properties, which are live and dead aboveground biomass (AGB).

AGB is a major component of ecosystem functioning, being a temporary pool of fixed carbon, and providing the basis for agricultural exploitation by cattle-grazing and mowing. AGB of salt marshes is also of major importance for nature conservation, e.g. by providing feeding ground for many bird species, especially for migratory geese (Bakker et al. 2005, Blew et al. 2005). Various studies have focused on the production of biomass in salt marshes (Janiesch 1991, Bakker et al. 1993). However, to our knowledge there is still lack in the study of linkages between environmental conditions, plant functional traits and biomass production. Particularly, studies that merge the field of botanical scaling (Enquist and Niklas 2002) with the scaling from environment through communities and ecosystem properties (Naeem and Wright 2003, Suding et al. 2008) are missing. Salt marshes are particularly suited for this purpose because their plant species pool is relatively small. Therefore, a single study can quantify species' biomass allocation patterns, environmental conditions and ecosystem properties.

The main emphases of this paper are the assignment of the effect-response framework to salt marsh habitats as well as the exploration of allometric and non-allometric relationships among plant traits in order to understand the functional relationships that determine ecosystem properties. We address the questions of how species traits respond to environmental conditions of salt marshes and devise trait-interactions and trade-offs between functional traits of salt marsh plants. Finally, we point out effect traits that determine production of standing biomass in salt marshes. We assume that (i) trade-offs and allometries in biomass allocation determine the relationship between environmental conditions and community standing biomass in the response-effect framework; (ii) resources and abiotic stress factors such as salt and inundation are the primary agents determining response traits (Lavorel and Garnier 2002); and (iii) we expect a small set of keystone traits to define ecosystem property.

To specify these assumptions, we constructed a conceptual model based on *a priori* knowledge, which was tested against a dependence model using path analysis and structural equation modelling (Grace and Pugesek 1997). These techniques allow for the evaluation of alternative models, the partitioning of direct and indirect effects and the test for model-fitness (Mitchell 1992, Grace and Pugesek 1998). Furthermore, to explore how scaling relationships translate from the species to the community level, we determined the bivariate scaling relationships between biomass allocated to roots and rhizomes, leaves, stems and diaspores at level of individual plants.

The hypothesised model: Linkages between traits

We begin the hypothesised model with linkages between plant traits because they determine the biological functioning of plants (Fig. 8-1a). Thereafter, we connect the traits with the environmental variables (Fig. 8-1b) and the biomass variables (Fig. 8-1c).

Allocation theory assumes that organisms have a limited supply of resources which they have to allocate amongst their functions, broadly defined as growth, maintenance, storage and reproduction (Bazzaz 1997). Two different perspectives can be found in the literature: 'partitioning' and 'allometry' (Weiner et al. 2009). The partitioning perspective is interested in shifts in allocation patterns among plant structures and functions on environmental gradients, expressed in ratios such as root:shoot ratio or reproductive effort (reproductive biomass/total biomass, Obeso 2002). By definition, ratio-based biomass partitioning measures are size-independent, i.e. a large plant may exhibit the same root:shoot ratio as a small plant. On the other hand, the allometry perspective emphasises variation in size and states that scaling relations among standing leaf, stem and root biomass are generally positive, e.g. that root biomass increases with leaf and stem biomass (Enquist and Niklas 2002). Here, we are interested in understanding the variation of a size-dependent ecosystem variable, community standing biomass, in relation to plant biomass allocation patterns and environmental conditions. Therefore, we adopted the allometry perspective rather than the partitioning perspective and quantified the dry weight of leaves, stems, diaspores and below-ground organs (i.e. root and rhizome mass).

Leaves (and to a lesser degree stems) provide the plant with assimilation products, which is used by the plant to form new tissue. We thus expect a positive relationship between leaf biomass on the one hand and stem and diaspore biomass on the other hand (Fig. 8-1a, Enquist and Niklas 2002, Niklas and Enquist 2003, Weiner et al. 2009).

Species with high SLA (specific leaf area) also show high tissue nitrogen content (Wright et al. 2004). We therefore expect a negative relationship from C:N-ratio to SLA. We expect a positive relationship from SLA to leaf and stem biomass, as fast growing species produce higher leaf biomass and need sufficient supporting tissue.

Below-ground mass provides the basis for aboveground growth and (fine) roots provide the plant with water and nutrients, two essential prerequisites for photosynthesis, thus we draw a positive relationship between this trait and SLA, leaf and stem biomass. However, one can also expect a positive relationship between leaves and below-ground mass, as leaves provide the assimilation products for belowground growth.

Three general strategies to cope with salt stress have been identified, which are (i) exclusion of salt ions, (ii) dilution of cell sap and (iii) morphological adaptations to osmotic stress. The first strategy refers to species which feature structural and functional adaptations that reduce ion uptake through the root (Kinzel 1982), excrete salt ions via glands and bladders or continuously renew the basal leaves (Schirmer and Breckle 1982, Van Diggelen et al. 1986). Salt-diluting species accomplish dilution of cell sap via enlargement of mesophyll cells, which results in succulent growth (Kinzel 1982). Species that show morphological adaptations to osmotic stress roll in their leaves to decrease the transpiring area or sheathe their leaves with hair to attenuate incoming solar radiation (Rozema et al. 1985). Minimizing the transpiration rate thus diminishes the intake of saline water by the roots. Some species of the salt marsh show a combination of several of the strategies, whereas others only exhibit a single adaptation. For our model we classified the salt marsh species into two groups: those featuring excretion and dilution, and those featuring morphological adaptations to osmotic stress (MAOS). The latter applies to species of the upper salt marsh (*Elymus athericus*, *Festuca rubra* and *Artemisia maritima*), which do not show any other adaptations to salt stress. It could be assumed that a reduced leaf mass could gain an advantage to those species, as the transpiring area is reduced. We thus expect a negative relationship between MAOS and SLA as well as leaf biomass. However, we expect a positive relationship between C:N-ratio of the whole plant and MAOS, as the latter often comes along with xeromorphic tissue (Rozema et al. 1985).

The hypothesised model: Linkages between environment, traits, and standing biomass

We expect the environmental variables to be more or less correlated (Fig 8-1b). For instance, groundwater level and salinity ('Groundwater') should be strongly correlated with inundation frequency. By addressing these correlations (double-headed arrows), we also consider the indirect effects of a variable of interest on other variables through the correlations among the environmental variables.

Salinity and inundation frequency are considered the primary stress factors in salt marshes, from which salinity is regarded to as the 'master factor' (Rozema et al. 1985). Salinity affects biomass production and weight of the whole plant or roots and leaves of many salt marsh species (Lenssen et al. 1995, Egan and Ungar 2001).

Based on this background we assume the 'groundwater' variables level and salinity as well as inundation frequency to have negative effects on weight of leaves, stem, below-ground organs, MAOS and SLA.

Nutrient availability strongly affects primary productivity in salt marshes (Kiehl et al. 1997) and SLA is accounted as response to soil resource availability (Cunningham et al. 1999, Poorter and de Jong 1999). We assume nutrient availability (phosphorus, potassium and carbonate) to positively influence leaf, stem, diaspore, below-ground biomass and SLA. Olff (1997) has shown that nitrogen availability increases with silt content of salt marsh soils, which in reverse is a decrease with sand content. We therefore expect sand content to have negative influence on SLA, leaf and stem biomass, and positive on below-ground biomass and C:N-ratio. However, as increase in sand content also increases soil drainage and aeration, these suggested relationships could also be lessened.

Following De Deyn et al.(2008) plant traits that determine carbon and nutrient conservation are, amongst others, high C:N-ratio and long litter residence time. The slower the decomposition of plant material, the higher the accumulation of dead plant material, hence we expect C:N-ratio to explain dead biomass.

Aboveground biomass of the community (AGB) is the product of the biomass of leaves, stems and diaspores of the individual plants of the community. Therefore we expect stem biomass and leaf biomass of the species to explain live AGB (Fig. 8-1c). Garnier et al. (2001) attribute high SLA to rapid production of biomass, hence we expect a positive relationship between SLA and live AGB. Considering diaspore mass, cost of reproduction may be manifested in a decrease of survival (Biere 1995) at least when plants are subject to stress. We therefore expect a trade-off of resource allocation to either vegetative growth or reproduction (Obeso 2002). Finally, as dead AGB consist of formerly living biomass, we expect a positive relationship between live and dead AGB.

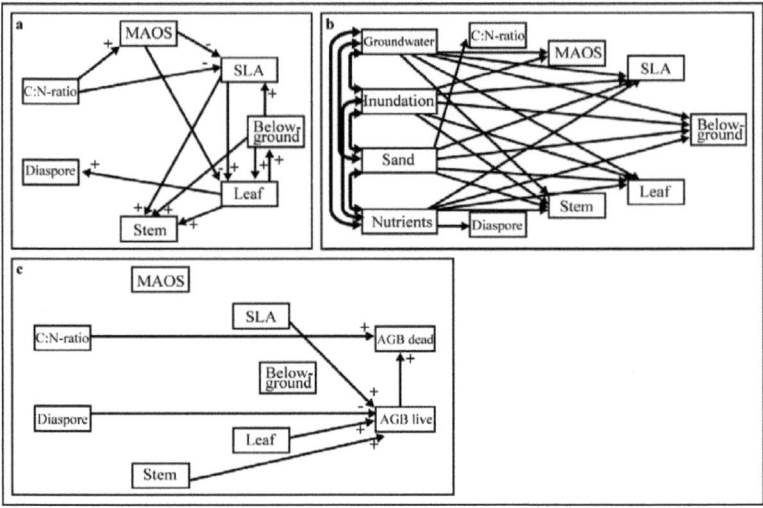

Figure 8-1: a) Hypothesised model of relationships between traits. Allometric relationships are indicated by '+', trade-offs between traits by '-', see also text above. b) Hypothetical model of relationships between environmental variables and plant functional traits. c) Hypothetical model of plant functional traits and community aboveground standing biomass. Single-headed arrows represent relationships; double-headed arrows represent free correlations. Residual error variables (e_x, effects of unexplained causes) were omitted from dependent variables. For names and abbreviations see Table 8-1.

8.2 Methods

Study area

The study was conducted at three study areas along the mainland coastline of Lower Saxony, Germany, and on the island of Mellum (52°43'N, 8°08E, 41 plots). On the mainland, field work was carried out in three study areas, these were: Leybucht (53°32'N, 7°07'E, 8 plots), Norderland (53°40'N, 7°19'E, 32 plots) and Jade Bight (53°26'N, 8°09'E, 32 plots). Mean annual temperature is about 9 °C and annual precipitation ranges from west to east from 770 – 830 mm/year (based on data from 1961 to 1990 for Norderney and Wilhelmshaven, Deutscher Wetterdienst 2009).

The foreland marshes of Lower Saxony extent over an area of 5430 ha (Bakker et al. 2005) and consist predominantly of clayey silt, loamy sand and loamy silt. They often developed through land reclamation (Pott 1995). The island of Mellum originated from sand deposits transported by currents along the coast from The Netherlands and Northwest Germany (Pott 1995). It now belongs to the core zone of the Wadden Sea National Park of Lower Saxony. Whereas some soils consist of silty clay and clayey loam, the predominant substrate is sand.

Sampling design

The selection of the sites was based on the stratification of the salt marshes of Northwest Germany regarding human-modified vs. natural conditions, silt vs. sand and unused vs. grazed or mown. Mellum was considered to be natural, with sandy soils and unused. The three mainland areas differed with respect to management. Within the study areas, the distribution of plots was based on random stratified sampling, with elevation as the stratification gradient.

Plots were sampled in summer 2007. At each of the 113 plots we collected soil samples from each soil horizon. Soil sampling was restricted to a depth of 30 cm due to upwelling groundwater. Soil samples were air dried, sieved through a 2 mm sieve and analysed for sand content (Ad-Hoc-AG Boden 2005) and calcium carbonate ($CaCO_3$, according to Scheibler in Schlichting et al. 1995). Soil available potassium and phosphorous were extracted with ammoniumlactate-acetic acid at pH 3 following Egnér et al. (1960) and analysed by AAS (Atomic Adsorption Spectroscopy) and CFA (Continous Flow Analyser, Murphy and Riley 1962), respectively. Groundwater level and salinity were measured from May to September 2007. Looped drainage pipes allowing water inflow were buried vertically in the ground at each survey plot. Groundwater level was recorded biweekly at low tide in the drainage pipes. Data were revised by regression analysis to generate mean values including high tide events, see Statistical Analysis. Salinity of groundwater was recorded by use of a conductivity measurement device ('WTWpH/Cond 340i/SET' with measuring electrode 'Tetracon 325'). Times series of both groundwater level and salinity were aggregated to their mean between May and September 2007.

Data loggers ('diver', ecoTech, Pegel-Datenlogger PDLA) were installed at 18 plots of different elevations to record inundation frequency. They were positioned at the bottom of the drainage pipes and recorded the salient water column at an hourly basis. Additional four data loggers recorded the pressure of the surrounding air, which was used to calculate the relative pressure of water accumulating in the pipe. Almost all abiotic and trait variables differed strongly between mainland and island salt marshes, see Table 8-1. Thus, mainland and island marshes were considered different habitats for this study.

Table 8-1: Environmental, trait and ecosystem property variables used for this study. Carbonate, potassium and phosphorous were aggregated by PCA-analysis; sample scores were named 'Nutrients'. The same was done for groundwater level and salinity; sample scores were named 'Groundwater'. Diaspore, stem, leaves and below-ground biomass value were log-transformed prior to analyses. Mean values and standard deviation (SD) for all variables on mainland and island plots are given in the last columns.

Environmental variables	Abbreviation	Unit	Transf.	Mainland Mean	Mainland SD	Island Mean	Island SD
Carbonate		t ha^{-1}	None	207.15	66.57	18.81	22.64
Potassium	Nutrients	kg ha^{-1}	None	1,858.55	607.04	853.75	333.98
Phosphorous		kg ha^{-1}	None	184.49	57.82	100.42	56.05
Groundwater level	Groundwater	cm	None	-30.97	12.52	-21.41	16.44
Groundwater salinity		PSU	None	22.47	4.75	24.87	3.36
Inundation frequency	Inundation	hours	None	172.82	273.97	242.39	297.94
Soil sand content	Sand	%	None	44.31	20.89	73.26	20.31

Trait variables	Abbreviation	Unit	Transf.	Mainland Mean	Mainland SD	Island Mean	Island SD
Dry weight of							
diaspore unit	Diaspore	g	log	1.79	1.20	0.58	0.44
stems	Stem	g	log	7.78	8.69	1.99	1.79
leaves	Leaves	g	log	1.27	0.62	1.88	1.61
below-ground organs	Below-ground	g	log	1.50	1.11	4.72	4.92
C:N-ratio of whole plant	C/N-ratio	None	None	37.98	7.56	35.54	5.51
Specific leaf area	SLA	mm²/mg	None	12.06	2.79	9.40	2.52
Morpholgical adapation to osmotic stress	MAOS	1-Yes/ 0-No	None	0.37	0.33	0.25	0.38

Ecosystem property variables	Abbreviation	Unit	Transf.	Mainland Mean	Mainland SD	Island Mean	Island SD
Aboveground biomass live	AGB live	g m^{-2}	None	466.99	191.22	229.59	164.95
Aboveground biomass dead	AGB dead	g m^{-2}	None	182.64	104.98	94.21	91.35

Aboveground biomass was sampled in August, which equals peak biomass (De Leeuw et al. 1990). Samples were collected on an area of 0.5 m² on each survey site and sorted according to live and dead plant material. Samples were oven dried at 70 °C for 72 hours and weighted. Species composition and abundance was evaluated by frequency analysis using a 1 x 1 m frame subdivided into 100 grids of 0.1 x 0.1 m. Nomenclature followed Flora Europeae in SynBioSys Species Checklist (2010).

Plant functional traits were sampled for 242 individuals, i.e. approximately 10 individuals per species. These individuals were selected from different plots covering the total range of species' occurrences within the environmental space. For the measurement of biomass allocation, plants were collected at the peak of their generative stage, i.e. when seeds were ripened but not yet shed. These data were used for the calculation of the community weighted means of stem, leaf, diaspore and below-ground biomasses (see below). Additionally, we collected 112 individuals from 14 mainland species at the peak of their vegetative stage, i.e. before flowering. Plants were dug out, roots and rhizomes were cleaned of soil material by rinsing off the soil substrate and roots of different individuals were carefully separated using tweezers. Plant material was subsequently oven dried at 70 °C for 72 hours. Leaves, stems, diaspores and below-ground organs (roots and rhizomes) were weighted after drying. C:N-content was measured after grinding the plant material in a planetary mill for about 2 to10 minutes at 300-400 revolutions ('pulverisette 7', Fritsch). Each sample was then dried at 105 °C for 4-5 hours. 2-3 mg of material was filled into tin-tubes (0.1 mg precision balance CP 225 D, Sartorius; tin capsules for solids, Säntis Analytical) and analysed using CHNS-Analyser Flash EA (Thermo Electron Corporation) following Allen (1989).

SLA (Specific leaf area) was calculated as the ratio of leaf area to leaf dry mass (mm²/mg) following Knevel et al.(2005). The trait morphological adaptation to osmotic stress (MAOS) was evaluated by literature research with subsequent assignment of either the existence (1) or the non-existence (0) of this attribute to a species.

Statistical Analysis

Groundwater level

At all plots, groundwater level was recorded biweekly during low tide and thus contained no information about variation due to tidal elevation. To adjust this lack of information, a regression was conducted with paired values of the hourly data from the data loggers at a subset of 18 plots and the biweekly data of the groundwater level measured at these plots. Mean values of both groundwater and data logger over time were generated for each of these plots and used for the linear regression analysis. The regression function was used to adjust values of mean ground water level of all other plots to include information about tidal effects.

PCA of environmental parameters

There were strong correlations between carbonate, potassium and phosphorous, as well as between groundwater level and groundwater salinity. Therefore, to reduce the number of variables in the model and using principal component analysis (PCA), carbonate, plant available potassium and phosphorous were aggregated to their first principal component called 'nutrients'. Likewise, groundwater level and salinity were aggregated to 'groundwater'. Analysis was carried out using the computer software R (The R Foundation for Statistical Computing 2008).

Bivariate scaling relationships at the individual level

Standardised Major Axis regression (SMA, Warton et al. 2006) was used to describe bivariate relationships between traits at the individual level. Trait variables are biologically interdependent and have similar variation associated to them due to measurement error. In this case, ordinary least squares regression is considered inappropriate (Niklas 1994a), because it keeps the x-values fixed, and finds the line which minimizes the squared errors in the y-values, whereas SMA tries to minimize both the x- and the y-errors. SMA regression was done with the software PAST (version 1.94b, Hammer et al. 2001).

Community weighted mean

According to Grime (1998), the extent to which a plant species influences ecosystem function is likely to be derivable from its contribution to total plant biomass which is indicated by its abundance. To predict AGB from traits, we therefore used community weighted means (Garnier et al. 2007). The mean value of each trait of every species was weighted by the abundance of the species on each plot and averaged by the total abundance of every species on each plot. A list of all variables used in the path analysis is given in Table 8-1.

Path analysis and Structural Equation Modelling (SEM):

Path analysis was used to quantify the relationships among variables of our a priori model, i.e. the relations between abiotic conditions, plant functional traits and ecosystem functioning (standing biomass). As we only used observed variables in our study, our model is manifest (McCune and Grace 2002). Thus, we used path analysis for partitioning the correlations among variables and to measure both direct and indirect effects on response variables (Kingsolver and Schemske 1991). SEM was used to test overall model fitness (Grace and Pugesek 1998).

The hypothetical model (Fig. 8-1) generates an expected covariance structure, which is compared to the actual covariance matrix. Standardized coefficients describe the strength of the relationship. Relationships between variables are either uni-directionally causal (indicated by a straight, single-headed arrow) or unanalyzed correlations (indicated by a curved, double-headed arrow). Indirect pathways between variables involve intermediary variables. Direct pathways are the value of path coefficients, which are the standardized partial regression coefficients of the displayed arrow that directly connect two variables (Grace 2006).

SEM allows testing the hypothesis that the measured covariance structure adequately describes the expected covariance structure. This is done by means of maximum likelihood estimates which generate a test statistic that is distributed approximately as χ^2 (Mitchell 1992). Good fit of the hypothesized model to the data will result in low χ^2/df and non-significant p-value. Goodness-of-fit-index (GFI) is a measure of the relative amount of variance and covariance that the model allows for. Another index of fit, the root mean square error of approximation (RMSEA, Browne and Cudeck 1993) assesses closeness to fit. Good models have RMSEA < 0.05, $p > 0.05$, χ^2/df-values < 2, and GFI > 0.9 (Backhaus et al. 2003). Path analysis and SEM was done using Amos 16.0.1 (Arbuckle 2007).

8.3 Results

Environmental and ecosystem property variables

A comparison between mainland and island revealed significant differences between all environmental parameters and biomass values at $p > 0.05$ (Table 8-1). Soils of mainland marshes were richer in nutrient availability and contained less sand than island marshes. AGB was higher in mainland marshes than in island marshes. High standard deviations resulted partly from the orientation of the plots along the elevation gradient, for example plots in the pioneer zone got inundated about 100 to 600 hours during sampling period, whereas plots in the upper marsh were only inundated about 3 to 100 hours. As there were more plots in the lower salt marshes of the island than of the mainland, mean inundation frequency was higher in island marshes.

Trait variables

Sixteen different species were sampled which represented about 90% of the regional salt marsh species pool. Four were grasses (*Spartina anglica*, *Puccinellia maritima*, *Festuca rubra* and *Elymus athericus*) and two were perennial sub-shrubs which develop woody stems (*Artemisia maritima* and *Atriplex portulacoides*). The remaining species were annual and perennial herbs.

Trait values for the community of mainland and island differed significantly for diaspore, stem and below-ground weight, as well as for MAOS and SLA at p > 0.05 (Table 8-1). Only C:N-ratio and leaf biomass resulted in no significant differences. High standard deviations resulted from different trait expressions along the elevation gradient, e.g. *Aster tripolium* of the lower marsh produced higher stem biomass (mean of 41.9 g for mainland) than *Elymus athericus* from the higher marsh (mean of 0.32 g for mainland).

PCA of environmental parameters

Correlation coefficients for nutrients were high (potassium – carbonate: 0.67, potassium – phosphorous: 0.53, and carbonate – phosphorous: 0.65) and mean groundwater level and salinity also showed a high correlation (0.64). All correlation coefficients were significant at the p > 0.01 level.

Explained variance for nutrients (i.e. the 1st principal component of potassium, phosphorous, and carbonate) and groundwater (the 1st principal component of groundwater level and salinity) are high for the first axis (0.74 and 0.82) and accordingly lower for the second (0.16 and 0.18, Fig. 8-2). PCA sample scores of nutrients and groundwater were used for further path analysis.

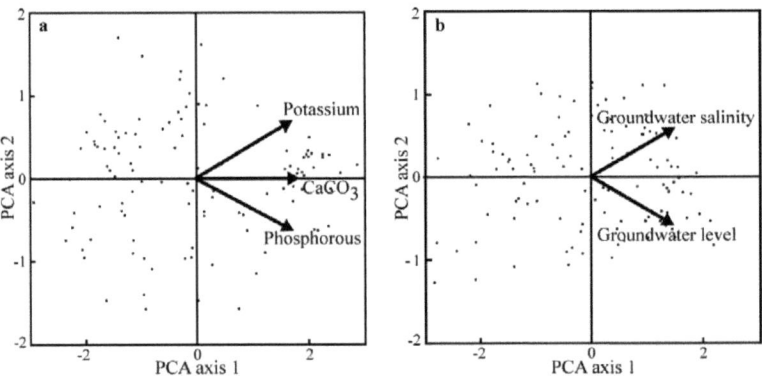

Figure 8-2: PCA of environmental parameters using potassium, carbonate and phosphorous (A) and groundwater salinity and groundwater level (B).

Biomass partitioning at the individual level

At the individual level, log-log relationships between pairs of biomasses of organs were close to isometric (Table 8-2). Only the relationships between below-ground biomass (roots and rhizomes) and leaf as well as diaspore mass were significantly different from isometry, for the vegetative and the generative stage. For the generative stage, r² values were lower than for the vegetative stage.

This may be attributed to the fact that leaves of some individuals were already close to senescence at the peak of the generative stage.

Table 8-2: Statistics of pairwise reduced major axis regression analyses (RMA) of standing leaf, stem, below-ground and diaspore biomass (M_L, M_S, M_R and M_D, respectively). M_R comprises root and rhizome masses. The slope of the regression is denoted by α_{SMA} and the intercept by β_{SMA}. P ($\alpha = 1$) shows the probability that slope α of the regression is isometric. In all cases, $P < 0.0001$.

Y_2	vs.	Y_1	α_{SMA}	SE	β_{SMA}	SE	95%CI	r^2	n	$P(\alpha=1)$
Peak vegetative stage										
M_L	vs.	M_S	0.94	0.05	-0.079	0.040	0.85-1.03	0.73	112	0.217
M_S	vs.	M_R	0.96	0.06	0.263	0.061	0.85-1.08	0.57	112	0.528
M_L	vs.	M_R	0.90	0.04	0.167	0.046	0.80-0.99	0.72	112	0.042
Peak generative stage										
M_L	vs.	M_S	0.98	0.05	-0.393	0.042	0.89-1.09	0.36	242	0.742
M_S	vs.	M_R	0.91	0.05	0.142	0.040	0.82-1.01	0.35	242	0.060
M_L	vs.	M_R	0.89	0.04	-0.252	0.038	0.83-0.95	0.52	242	0.009
M_D	vs.	M_R	0.86	0.04	-0.277	0.037	0.79-0.94	0.51	242	0.001
M_D	vs.	M_L	0.97	0.05	-0.032	0.047	0.87-1.07	0.37	242	0.570
M_D	vs.	M_S	0.95	0.03	-0.413	0.025	0.90-1.01	0.75	242	0.147

Path analysis and Structural Equation Modelling (SEM):

Mardia's coefficient (Mardia 1970, 1974) suggested a non-normal distribution of the data of the initial model. To obtain multivariate normality, 9 sample points were deleted, using Mahalanobis distance as indicator. Modification indices suggested that adding several paths might substantially improve model fit (inundation → C:N-ratio; groundwater → C:N-ratio; nutrients → MAOS; SLA → diaspore, dead AGB; stem → diaspore, dead AGB; C:N-ratio → diaspore; below-ground → diaspore, MAOS, live AGB, dead AGB).

Several paths were nonessential to the model, and hence were deleted (inundation ↔ sand; inundation ↔ nutrients; groundwater → below-ground; inundation → below-ground, → leaf, → stem, → diaspore, → MAOS, → SLA; sand → C:N-ratio, → below-ground, → SLA; nutrients → leaf; leaf → below-ground, → diaspore, → AGB live; below-ground → stem, → leaf; diaspore → live AGB; MAOS → SLA; live AGB; C:N-ratio → below-ground).

The changes made to the initial model led to a stable model that only included essential pathways, which were all significant at $p < 0.05$.

The resulting model showed good consistency with the data ($\chi^2/df = 1.022$, $p = 0.431$, GFI = 0.947, RMSEA = 0.015) and explained 45 % of the variation in live AGB and 48 % in dead AGB.

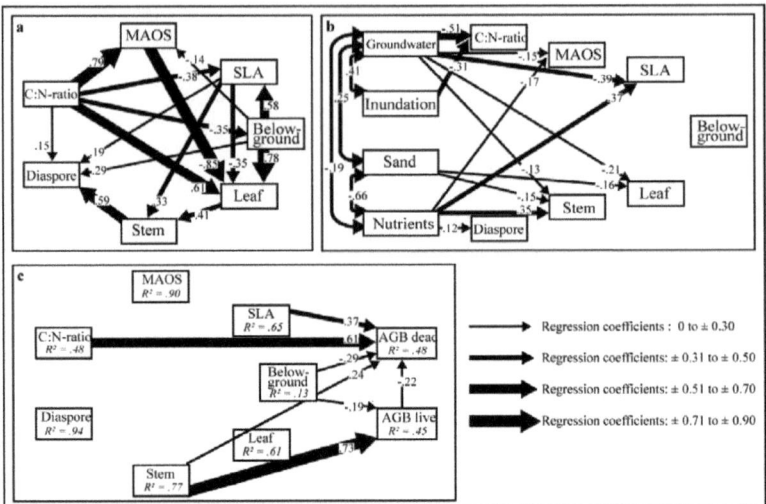

Figure 8-3: Final models derived from initial models in Figure 1. Names and abbreviations of observed variables follow Table 8-1. Path coefficients between variables are standardized partial regression coefficients of direct effects, for total and indirect effects see Table 8-3. Arrow widths are proportional to the standardized path coefficient. Variances explained by the model (R^2) are given under the variable names. Residual errors were omitted. All pathways are significant at $p < 0.05$. a: trait-trait relationships, b: environment–trait relationships; c: trait relationships to Aboveground standing biomass live and dead.

The structural equation model revealed both direct and indirect effects between environmental conditions, traits and aboveground biomass. Direct effects are visualized in Fig. 8-3a-c. Indirect effects occur if two variables are connected through paths to and from a third variable (see Methods). Total effects are calculated from the combined direct and indirect effects (Table 8-3). Here, we concentrate on relevant total effects, i.e., those >0.3 in Table 8-3.

Table 8-3: Standardized total, direct and indirect effects of factors that influence aboveground biomass production. Path coefficients between variables are standardized partial regression coefficients. Total effects were calculated by adding direct and indirect effects. All effects were significant at p < 0.05.

	Groundwater			Inundation			Sand			Nutrients			AGB live		
	Tot.	Dir.	Ind.	Tot.	Dir.	Ind.	Tot.	Dir.	Ind.	Tot.	Dir.	Ind.	Tot.	Dir.	Ind.
C:N-ratio	-.51	-.51		-.31	-.31										
Below-gr.	.18		.18	.11		.11									
SLA	-.09	-.39	.29	.18		.18				.37	.37				
MAOS	-.58	-.15	-.43	-.24		-.24	-.16	-.16		-.17	-.17				
Leaf	.15	-.21	.36	.06		.06	-.22	-.15	-.07	.01		.01			
Stem	-.11	-.13	.03	.08		.08	-.13		-.13	.48	.35	.13			
Diaspore	-.11		-.11	.07		.07				.48	.12	.36			
AGB live	-.11		-.11	.04		.04	-.15		-.15	.35		.35			
AGB dead	-.39		-.39	-.14		-.14	-.02		-.02	.17		.17	-.22	-.22	

	C:N-ratio			Below-ground			SLA			MAOS			Leaf			Stem		
	Tot.	Dir.	Ind.	Tot.	Dir.	Ind.	Tot.	Dir.	Ind.	Tot.	Dir.	Ind.	Tot.	Dir.	Ind.	Tot.	Dir.	Ind.
Below-gr.	-.35	-.35																
SLA	-.58	-.38	-.20	.58	.58													
MAOS	.84	.79	.05	-.14	-.14													
Leaf	-.18	.61	-.79	.70	.78	-.08				-.85	-.85							
Stem	-.26		-.26	.48		.48	-.35	-.35		-.35		-.35	.41	.41				
Diaspore	-.22	.15	-.37	.68	.29	.39	-.19	.33	-.14	-.21		-.21	.24		.24			
AGB live	-.12		-.12	.16	-.19	.35	-.30	.19	-.11	-.25		-.25	.29		.29	.59	.59	
AGB dead	-.45	.61	-.15	-.14	-.29	.29	-.38	.37	-.01	-.03		-.03	.03		.03	.08	.24	-.16

Trait-trait linkages

The strongest allometric relationships were between below-ground organs and leaf mass, stem mass, diaspore mass and SLA and between stem and diaspore mass and leaf and stem mass (total effects Table 8-3, direct effects Fig. 8-3a). C:N-ratio showed strong allometric relationship to MAOS and the strongest trade-off was between MAOS and leaf mass, i.e. in the upper marsh tissue of plants is rich in carbon and allocation towards leaves is low.

C:N-ratio showed a strong trade-off to SLA and below-ground mass, which means that, considering the relationships already mentioned, upper marsh species show low SLA values and low allocation to below-ground organs.

Response traits

Sand was negatively correlated to nutrients and positively to groundwater, whereas groundwater was positively correlated to inundation frequency (Fig. 8-3b). This indicates that the lower parts of the salt marsh with high inundation frequency are less rich in nutrients than the upper parts.

The traits responding most strongly were SLA, stem and diaspore mass with positive responses to nutrients (total effects Table 8-3, direct effects Fig. 8-3b) and C:N-ratio and MAOS with strong negative responses to groundwater and inundation frequency (the latter with response from C:N-ratio only). There were no relevant responses to sand content.

Effect traits

Stem biomass together with leaf mass had the strongest positive effects on standing live aboveground biomass, however, there was no direct relationship between leaf mass and AGB live (Fig. 8-3c). Leaves exerted influence on standing live biomass through an allometric relationship with stem mass. C:N-ratio and SLA the strongest negative effects on AGB dead. Hence, stem biomass and C:N-ratio were both response and effect traits.

Indirect effects of environmental variables on standing biomass

By their effect on the traits, groundwater had a negative indirect effect on aboveground dead biomass and nutrients a positive, but weak indirect effect on aboveground live biomass.

8.4 Discussion

Lavorel & Garnier (2002) referred to linkages between environment, biodiversity and ecosystem properties via response and effect traits as the 'holy grail' of functional ecology. Path analysis and structural equation modelling allowed us to quantify these linkages for a single ecosystem property:

above-ground biomass of salt marshes. In particular, we were able to demonstrate that the allometric biomass allocation pattern at the species level translates into the community level. Through the allometric relationship among stem, leaf, below-ground and diaspore mass, a single response and effect trait, stem biomass, drives the relationship between soil nutrients and aboveground live community biomass. On the other hand, whole-plant C:N-ratio and SLA were the main response and effect traits driving the relationship between aboveground dead community biomass and groundwater level and salinity and nutrients.

Scaling relationships in plant biomass allocation at the species level

To understand whether the SEM relationships among plant organ biomasses at the community level are consistent with those at the species level, we investigated interspecific allometric scaling relationships at the vegetative and generative stage. Most of these relationships were close to isometry, whereas, according to allometric theory, they should scale to the three-quarter power, except for the relationship among root and stem mass which should be isometric (Enquist and Niklas 2002, McCarthy and Enquist 2007). However, the three-quarter rule was empirically validated mainly for tree species, whereas deviations from this rule towards isometry were found for juvenile and herbaceous plants (Enquist and Niklas 2002, Niklas 2006). In our salt marsh species set, allometric relationships close to isometry were present at the vegetative and generative stage, and among vegetative and reproductive organs. In a recent review of allometric relationships among vegetative and reproductive biomass of herbaceous species, Weiner et al. (2009) found linear relationships close to isometry for the majority of the 76 species investigated. The scaling relationships at the species level suggest a common allometric strategy across all salt marsh species (Müller et al. 2000) implying that increase in the biomass of one organ is accompanied by a proportional increase in biomass of all other organs.

Relationships among plant traits at the community level

At the community level, i.e. when the mean organ biomass of all species present in a community weighted by their abundance was considered, the modelled relationships among stem, leaf, diaspore, root and rhizome mass were also allometric, as shown in Table 8-3. Corresponding to our hypothesised model, we found strong positive total effects of below-ground biomass on leaf, stem and reproductive biomass as well as of stem biomass on diaspore biomass. Total effects of leaf biomass on stem and diaspore mass were less pronounced. The path analysis and structural equation model explores these relationships in a multivariate rather than a bivariate way which has been rarely done so far (Shipley 2004).

Our expectation that leaf biomass would scale strongly negative with MAOS as a salt stress regulation mechanism was also confirmed. MAOS attributes to species of the upper salt marsh, which is a late successional stage in this habitat. Leaf mass proofed a good predictor for partitioning the salt regulation mechanisms between the upper and the lower salt marsh.

Trait responses to environmental conditions

Our final model illustrates that groundwater level and salinity as well as nutrient availability were the driving factors influencing functional traits whereas inundation frequency and sand content of the soil played minor
The relationship between inundation frequency and roles.

Contrary to the hypothesised model, there were no direct links between nutrients and leaf or belowground biomass. However, nutrients had a direct effect on SLA and on biomass allocation to stem. Additionally, sand content, being strongly negatively correlated to nutrients in this study and considered as a proxy for low nitrogen availability in salt marshes (Olff et al. 1997), had a negative total effect on stems. With increasing nutrients, plants invest more into their supporting stem tissue, thus seizing the favourable conditions to produce more vegetative tissue and to acquire dominance by occupying vertical space, and by that to compete for light (Tilman 1988, Westoby et al. 2002). In a greenhouse experiment, Kuijper et al. (2005) showed that the aboveground relative yield of *Elymus athericus*, a dominant plant of the high salt marsh, increased with nitrogen availability at the expense of *Festuca rubra*. *Elymus* also invested relatively more biomass in stem and root tissue and had larger shoot length than *Festuca*. This experiment confirms that changes in nutrient supply affect stem biomass allocation which results in changes in species abundances. Similar results were found in transplant experiments on Dutch salt marshes where increasing soil nutrient levels indicated by decreasing sand and increasing clay content favoured species with comparably high investment in stem tissue whereas rosette species decreased in abundance (Dormann et al. 2000).

Through the allometric relationship between stems and other plant organs, nutrients also indirectly favoured diaspore production (Niklas and Enquist 2003), thus increasing the chance of plants to recruit (Moles and Westoby 2006) and to explore new habitats via their propagules.

Groundwater exhibited greatest negative influence on C:N-ratio and MAOS, i.e. plants growing at high groundwater levels and salinities tend to stock relatively less carbon than nitrogen compared to those subjected to more balanced water conditions. This is in contrast to terrestrial conditions where plants growing in more stressful habitats are often characterised by higher C:N-ratios (De Deyn et al. 2008). However, many plants capable to withstand salt stress and frequent inundation synthesise nitrogen-requiring osmoprotectants such as proline, glycine betaine and other quaternary ammonium compounds to counter osmotic stress (Steward et al. 1979, Tarczynski et al. 1993). These osmotic solutes often make up a large part of the plant's nitrogen budget (Rozema et al. 1985).

groundwater level and salinity was not as straightforward as expected as the correlation between these two parameters was only moderate (0.41). We found that less frequently inundated higher salt marshes also display high groundwater salinities. In these cases, high salinity may arise from summer rain fall deficits (precipitation minus evapotranspiration), which causes decreased water content of the sediment and hypersaline conditions of the soil substrate (Leeuw et al. 1990). This may explain why C:N-ratio responded more strongly to groundwater level and salinity than to inundation frequency.

Keystone response and effect traits

Our model detected four traits explaining community standing biomass. These are C:N-ratio and SLA explaining dead biomass on the one hand and leaf and stem mass, which affect living biomass on the other hand. From these traits, C:N-ratio, SLA and stem mass were both effect and response traits.

Live biomass is explained by stem biomass because, based on the isometric allometry of all aboveground organs including leaf mass, an increase in stem biomass leads to a concurrent biomass increase in all other organs.

On the other hand, dead biomass is determined by the plant C:N-ratio with increasing carbon stocks leading to a higher dead fraction of the total biomass. We assume that the mechanism behind this link is an accumulation of litter due to reduced decomposition in the upper marsh. This assumption was confirmed by decomposition studies performed for a subset of the plots showing that the litter of plants exhibiting high C:N-ratio, particularly *Elymus athericus*, decomposed more slowly than plant litter with lower C:N-ratio (data not shown). Species from early successional stages tend to decompose more rapidly than those from more advances stages (Garnier et al. 2004, Kazakou et al. 2006). In salt marshes, *Elymus athericus* forms the successional climax stadium of the upper marsh (Bakker et al. 2003).

De Deyn (2008) showed that when soil resources are reduced, e.g. water, traits that drive carbon and nutrient conservation dominate, such as high C:N-ratio and longer litter residence time (Aerts and Chapin 2000). Our study showed that high groundwater level and salinity negatively influence C:N-ratio, i.e. high groundwater leads to low C:N-ratios, as the demand for nitrogen for osmoregulation increases at high levels of groundwater. We assume that plants of the upper marsh exhibit nitrogen demand for osmoregulation to a lesser degree than those of the lower marsh and of the pioneer zone. Thus, in the upper marsh, C:N-ratio is higher but also litter decomposition is reduced, which leads to the accumulation of dead biomass.

The second trait to explain dead biomass is SLA, though the effect on AGB dead was not as pronounced as the effect of C:N-ratio. SLA is higher in the lower marsh, which is expressed by the strong negative relationship to C:N-ratio. Fast-growing species from nutrient-rich sites show high SLA together with high tissue nutrient content and short lived leaves (Lavorel et al. 2007). We assume that species of the lower marsh show a high production of biomass, but as the turnover rate of this biomass is high the proportion of dead biomass increases. Dead plant material in the lower marsh can be washed away more easily by incoming water or spring tides than in the upper marsh, so that an accumulation of dead biomass in the upper parts of a salt marsh relative to the lower parts seems obvious. However, Bakker et al. (1993) estimated decomposition rates in salt marshes of The Netherlands and found that tissue composition (especially lignin content) was the most important factor determining rates of decomposition and environment heterogeneity (water, silt and salt content of the soil) was only the second important factor. This supports our assumption, that the weaker effect of SLA on dead biomass in comparison to C:N-ratio is a combination of high production of biomass of the lower marsh species on the one hand and higher losses due to decomposition.

Conclusions

The relationship between standing biomass and environmental conditions has often been addressed as a direct dependency. Here we emphasise the indirect link, i.e. plant traits respond to the environment and simultaneously affect aboveground standing biomass. Keystone response and effect traits were (i) stem biomass responding to available nutrients and affecting standing live biomass as well as (ii) plant C:N-ratio and SLA responding to groundwater level and salinity, and nutrients and affecting standing dead biomass. Stem biomass drives standing live biomass via an underlying isometric allocation of biomass to all other plant organs, including leaves, conforming to allometric theory. Likewise, the relevance of whole plant C:N-ratio in determining dead biomass has been confirmed for many different ecosystems, with SLA being its counterpart for the lower marsh. Therefore, these traits can be considered general keystone markers of species' functioning in determining the relationship between environment and ecosystem properties. The most relevant salt marsh specific result was the inverse relationship between C:N-ratio and environmental stress, as compared to most other terrestrial ecosystems. Lower C:N-ratio in plants with increasing salt stress is apparently due to higher nitrogen investments to synthesise osmoprotectants under salt stress.

Acknowledgements

We thank the administration of the National park 'Niedersächsisches Wattenmeer' and 'Mellumrat e.V.' for their support during field work. Many thanks to H. Timmermann and J. Spalke for contributing their data, to S. Andratschke for field assistance, to G. Schweiffarth and M. Heckroth for supporting our work at Mellum, and to E. Garnier for valuable comments on previous versions of the manuscript. This study was conducted as part of the TREIBSEL project and was financially supported by the 'II. Oldenburgischer Deichband' and the 'Wasserverbandstag e.V.' (NWS 10/05).

9 Plant trait responses to the environment and effects on ecosystem properties in salt marshes

Vanessa Minden and Michael Kleyer,
in preparation

Abstract

The assignment of different traits to ecosystem properties and to the species from which they arise will contribute to the understanding of loss of biodiversity, particularly against the background of global change and human induced habitat destruction. In our study, we explored the responses of plant functional traits of salt marsh species on the community level to the predominant environmental parameters and examined their effects on ecosystem properties (i.e. aboveground biomass (AGB), ANPP (Above Net Primary Productivity), decomposition and species richness).

We used path analysis to evaluate relationships between environmental parameters, functional traits and ecosystem properties and estimated model fitness by structural equation modeling (SEM).

Keystone response traits were belowground dry matter (BDM) responding to groundwater level and salinity and LSP ('leaf and stem properties' aggregated from SLA (Specific Leaf Area), LDMC (Leaf Dry Matter Content), SDMC (Stem Dry Matter Content) and SSD (Specific Stem Density)) responding to inundation frequency. The different ecosystem properties were affected from various plant traits, from which LSP and BDM were most relevant. However, our study clearly demonstrates that different ecosystem properties are affected by different plant traits, and the interaction of these traits contributes to the successful functioning of the ecosystem. We were also able to distinguish between effect traits of the pioneer zone, lower and higher marsh. Aboveground biomass (AGB) was most strongly affected by trait expressions of upper marsh species, as was decomposition. ANPP and species richness were most responsive to trait-combinations from upper and lower marsh species as well as pioneer zone plants. This underlines the importance of functional diversity and the significance of species conservation for maintaining ecosystem properties.

Keywords: effect trait, response trait, path analysis, structural equation modelling, salt marshes

9.1 Introduction

The term 'ecosystem function' is widely used to sum up various ecosystem properties, for instance biogeochemical pools and fluxes such as decomposition and primary productivity, without any reference to benefits for human life (Hooper et al. 2005). These ecosystem properties are strongly influenced by the functional characteristics of species forming a community (Díaz et al. 2006, Díaz et al. 2007). However, traits explaining one ecosystem property might not necessarily be relevant for another (Hooper et al. 2005). Therefore, identifying key traits relevant for ecosystem properties might both be beneficial for understanding ecosystem processes and helpful in conceiving the impacts of loss of species diversity (Chapin et al. 1997).

The development of plant traits which influence ecosystem properties depends on the environment, as plant species are sorted to different ecosystems via the response of their traits to the environmental conditions of the ecosystems (Díaz and Cabido 2001). The combined analysis of trait responses to the environment and their effects on ecosystem properties has been recently proposed by e.g. Suding et al. (2008) but rarely been applied in real landscapes (Pakeman 2004).

In this paper we present key traits explaining various ecosystem properties of salt marshes in northwest Germany and discuss overlaps in different trait–ecosystem property relationships. We constructed chains between environmental parameters exerting influences on functional traits (response traits), which in turn affect (effect traits) ecosystem properties.

Salt marsh plants are under constant influence of the tidal regime and are subject to changing geomorphological, physical and biological processes (Bakker et al. 2005). This results in a relatively small species pool, which redounded to our advantage of being able to collect plant traits from almost all species occurring in the habitat, both dominant and subordinate. In our study we analyse the response of the whole community on environmental influences and determine the effect of the community on various ecosystem properties.

The ecosystem properties chosen for this study are key components of carbon and nutrient cycles. These are: aboveground biomass (AGB), aboveground net primary productivity (ANPP), rates of decomposition of different time intervals and species richness, respectively.

The aim of our study was to point out how traits of salt marsh species respond to environmental conditions and how they affect ecosystem properties. By testing the effects on various ecosystem properties, we were able to directly evaluate recurrent patterns of effect traits, i.e. if different ecosystem properties are explained by the same plant traits or if different traits are relevant for different ecosystem properties.

We constructed a hypothesised model based on *a priori* knowledge, which was tested against a dependence model using path analysis (Grace and Pugesek 1998). Subsequent structural equation modelling allowed for evaluation of model fitness (Backhaus et al. 2003). The final model revealed direct effects between the various parameters, as well as indirect and total effects between all parameters. We only assumed effects of the environment on traits which in turn affect vegetation-based ecosystem properties, and included no direct paths between environmental factors and ecosystem properties. However, the indirect effects illustrate the relationships between environment and ecosystem properties as mediated by the species´ traits.

The hypothesised models: Trait-trait relationships

We started our conceptual model with trait-trait relationships (Fig. 9-1a), because allometric relationships and trade-offs between plant traits determine their influence on ecosystem properties. Subsequently, we linked these relationships with the environmental parameters and as a last step, with the various ecosystem properties (Fig. 9-1b and c-g). We chose a set of four species-specific traits and trait combinations for our models, which were aboveground and belowground mass (ADM and BDM), leaf and stem properties (LSP) and canopy height.

Aboveground dry mass (ADM) and belowground dry mass (BDM) of a species were considered separate traits for the analysis, as both of them accomplish different plant functions and show different effects on the ecosystem. Aboveground biomass is accountable for light interception and transpiration and provides feeding ground for e.g. migratory birds (Bakker et al. 2005), whereas belowground biomass captures soil resources and provides the plant with water and nutrients.

LSP consisted of the traits specific leaf area (SLA), specific stem density (SSD) and dry matter of leaves (LDMC) and stems (SDMC), which were aggregated via principal component analysis. The PCA scores for the species increased with decreasing SLA and increasing SSD, LDMC and SDMC (see results). These traits are components of the leaf and the plant economics spectrum and represent a trade-off between fast acquisition and turnover of carbon and nutrients and carbon/nutrient retention (Wright et al. 2004, Freschet et al. 2010). In other terrestrial plants, SLA-values show high correlation to leaf-N (see Freschet et al. 2010). However, salt marsh plants also use nitrogen for osmoregulation (production of compatible osmotic solutes, Rozema et al. 1985), which is an adaptation to salt stress. This led to only poor correlation of the two traits ($r=0.37$) and the omittance of leaf-N from LSP.

Canopy height at maturity is related to competitive ability and correlates allometrically with aboveground biomass (Westoby et al. 2002, Cornelissen et al. 2003). The ability of a plant to compete successfully for light is enhanced by the possibility to position its leaves in optimal arrangement in the canopy.

We expected negative relationships of LSP with canopy height, above- and belowground dry mass, because leaf and stem dry matter content correlate negatively with potential relative growth rate and fast production of biomass (Fig. 1a, Poorter and Remkes 1990, Cornelissen et al. 1996). On the other hand, plants investing in their aboveground shoot tend to have higher canopies, which is why we expected aboveground dry mass and canopy height to be positively related. Belowground organs provide the plant with water and nutrients, which is essential for successful growth. One could expect a positive relationship between belowground and aboveground dry mass, as high mass of belowground organs provide successful aboveground growth (Enquist and Niklas 2002). However, one could also expect a trade-off between these traits when nutrient conserving species with large belowground storage organs allocate proportionately less aboveground mass. In the latter case, we would also expect a trade-off between belowground dry mass and canopy height, as nutrient conserving species are often low in stature (Grime 1974).

The hypothesised models: Linkages between environment and traits
We expected the environmental variables groundwater, inundation and nutrients to be correlated which is indicated by double-headed arrows (Fig. 9-1b, Grace 2006).

The major stress factors for salt marsh plants are salinity and inundation frequency, which affect biomass production and influence growth (Rozema et al. 1985). We therefore expected the 'groundwater' variables level and salinity as well as inundation frequency to negatively influence above- and belowground dry mass and canopy height. We assumed a negative relationship of these environmental parameters to SLA, and in reverse expected a positive relationship to LSP.

Nutrient availability is essential for successful growth and stress tolerance of salt marsh species (Rozema et al. 1985). We therefore expected the relationships between nutrients (phosphorous and carbonate, as well as sand content as proxy for low nitrogen content, Olff et al. 1997) and plant traits to be opposed to those of groundwater and inundation frequency.

The hypothesised models: Linkages between traits and ecosystem properties
The ecosystem properties chosen for this study are major components of species' dynamics and of carbon and nutrient cycling (Figure 9-1c-g). AGB (aboveground biomass) is the total clipped aboveground live and dead matter at peak season, whereas ANPP (aboveground net primary productivity) is the net carbon gain of the aboveground vegetation over a given period of time plus any losses to death, decomposition and herbivores (Chapin et al. 2002). These ecosystem properties are critical to all trophic levels whose existence depend on primary producers and are important components of the global carbon cycle (Scurlock et al. 2002). Decomposition on the other hand is

the breakdown of organic matter, which releases carbon and nutrients essential for new growth and development of living organisms. Species diversity influences the resilience and resistance of ecosystems to environmental change by containing higher phenotypic trait diversity and by that stabilizing the energy flow among trophic levels, decreasing the susceptibility of the ecosystem to invasion and reducing the spread of pathogens (Chapin et al. 2000).

Instead of making a selection of potential relationships between traits and ecosystem properties in the hypothesised models, we decided to permit unbiased combinations between all traits and the relevant ecosystem property. However, we expected various relationships to be significant from the outset.

AGB is high in the upper marsh, which is a result of the low rate of decomposition and the high amount of accumulating dead biomass (Bakker et al. 2003). We expected a positive relationship between LSP and AGB, as the long litter residence time in the upper marsh originates from high dry matter content of the plants' tissue (Minden and Kleyer unpubl.). We also expected a positive relationship between aboveground dry mass (species' trait) and AGB (the biomass of the community), as the latter is the product of the aboveground masses of all plants in the plot.

Poorter and de Jong (1999) found a positive correlation between SLA and ANPP. SLA is also positively related to relative growth rate, which results in high canopy height (Reich et al. 1997). We expected LSP to negatively explain, and canopy height to positively ANPP.

Environmental conditions and litter characteristics control decomposition rates. Species with low LDMC and high RGR produce litter that decomposes quickly (Kazakou et al. 2006). These traits are directly and indirectly aggregated in the variable LSP, which we expected to negatively explain decomposition.

We expected a positive relationship between LSP and species richness, because species with low LSP-values show high acquisition of carbon and nutrients and high growth rates (see e.g. Kazakou et al. 2006 for negative relationships between dry matter of leaves and relative growth rate). These species also show high competitive ability which leads to low total species richness (Grime 1974). Opposite to that, we expected belowground dry mass to positively explain species richness, as nutrient conserving species with high belowground mass often show low SLA and low relative growth rate, facilitating co-occurrence of many species (Poorter and Remkes 1990).

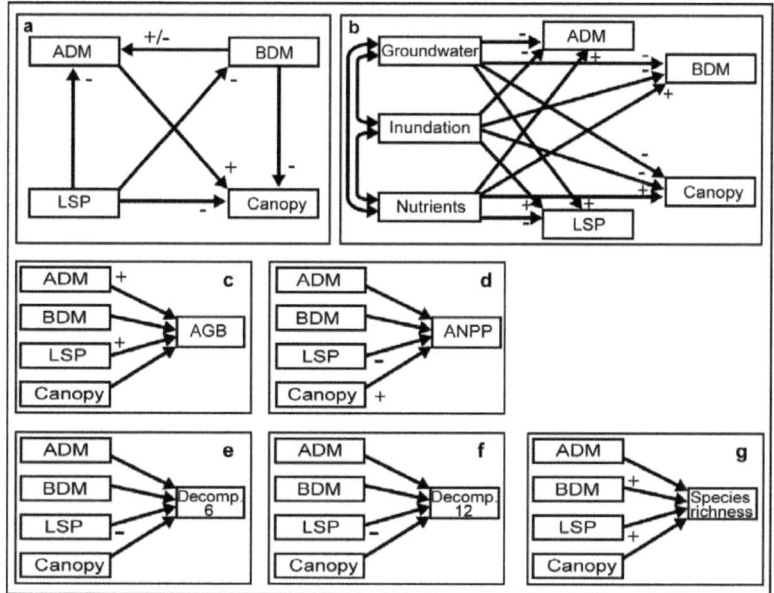

Figure 9-1: Structure of the hypothesised relationships between traits (a), environment and traits (b) and traits and ecosystem properties (c-g). Allometric relationships are indicated by '+', trade-offs by '-', no denotation indicates unbiased relationships. For names and abbreviations see Table 9-1. Single-headed arrows represent causal relationships; double-headed arrows represent free correlations.

9.2 Materials and Methods

Study area and sampling

Field work was carried out in three study areas along the coastline of Lower Saxony, Germany. Parts of the areas were subject to cattle grazing and mowing, whereas the major part was fallow land. These areas were Leybucht (53°32'N, 7°07'E), Norderland (53°40'N, 7°19'E), and Jade Bight (53°26'N, 8°09'E), see Figure 9-2. Climatic data from recent decades (1961 to 1990, for Norderney and Wilhelmshaven) showed a mean annual temperature of 9°C and a precipitation range from 770 mm – 830 mm/year (west to east, Deutscher Wetterdienst 2009).

The mainland marshes of Lower Saxony predominantly developed through land reclamation (Pott 1995) and extent over an area of 5 430 ha (Bakker et al. 2005). The prevailing type of soil is clayey silt, loamy sand and loamy silt.

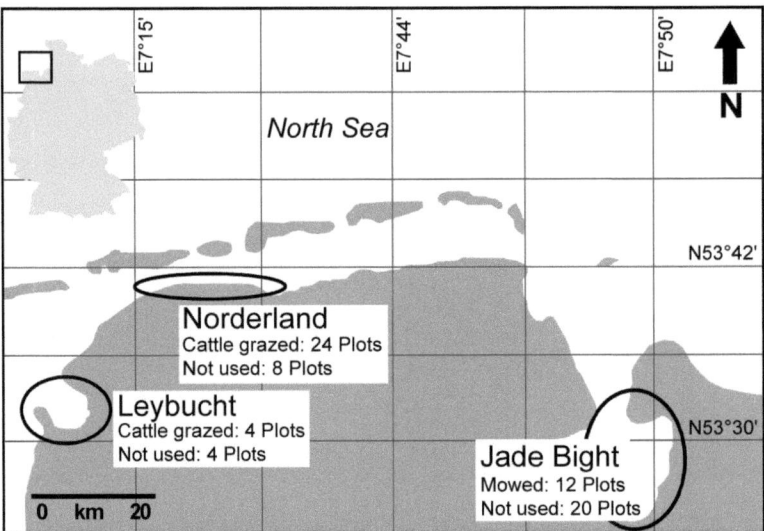

Figure 9-2: Location of study region in Germany (upper left corner, light grey) and position of three study areas along the coastline of northwest Germany with management regime and number of plots.

Within the study areas, distribution of plots was based on random stratified sampling, with elevation as stratification criterion. All 72 plots (4 m x 4 m) were sampled during the growing season 2007, as well as in spring and autumn 2008. Species composition and abundance was evaluated by frequency analysis using a 1 m x 1 m frame subdivided into 100 grids of 0.1 m x 0.1 m. Nomenclature of species followed Flora Europeae+ (SynBioSys Species Checklist 2010).

Environmental variables

Groundwater level and salinity were measured from May to September 2007. Perforated drainage pipes (6.5 cm diameter) were installed vertically in the ground at each plot. Groundwater level was recorded biweekly at low tide in the drainage pipes. Salinity of groundwater was recorded by use of a conductivity measurement device ('pH/Cond 340i/SET' with measuring electrode 'Tetracon 325'). Times series of both groundwater level and salinity were aggregated to their mean between May and September 2007.

Data loggers ('diver', ecoTech, Pegel-Datenlogger PDLA) were installed at 9 plots of different elevations to record inundation frequency and groundwater level at high tide. They were positioned at the bottom of the drainage pipes and recorded the salient water column each hour during the period of measurement. Additional three data loggers were placed nearby the study areas

(Leybucht, Norderland and Jade Bight) and recorded the pressure of the surrounding air, which was used to calculate the relative pressure of water accumulating in the pipes.

Soil sampling was restricted to a depth of 30 cm due to upwelling groundwater. Soil samples were air dried, sieved through a 2 mm sieve, and analysed for sand content (Ad-Hoc-AG Boden 2005) and calcium carbonate ($CaCO_3$, according to Scheibler in Schlichting et al. 1995). Plant available phosphorous was extracted with ammoniumlactate-acetic acid at pH 3 following Egnér et al. (1960) and analysed by CFA (Continous Flow Analyser, Murphy and Riley 1962). Table 9-1 summarizes all environmental variables.

Plant trait measurements

Plant functional traits were sampled for 14 different species, resulting in 112 individuals. These individuals were selected from different plots covering the total range of species' occurrences within the environmental space. Plants were dug out, roots and rhizomes were cleaned of soil material by rinsing off the soil substrate and roots of different individuals were carefully separated using tweezers. Plant material was subsequently oven dried at 70 °C for 72 hours. Leaves, stems, roots and rhizomes were weighted after drying (Table 9-1).

Aboveground dry mass (ADM) is the sum of the oven dry mass of leaves and stems, whereas belowground dry mass (BDM) consists of the dry mass of roots and rhizomes. SLA (specific leaf area) was calculated as the ratio of leaf area to leaf dry mass (mm^2/mg). LDMC (leaf dry matter content) is the ratio of dry leaf mass to fresh leaf mass (mg/g), the same counts for SDMC (stem dry matter content). Calculation of SLA, LDMC and SDMC followed Knevel et al. (2005). SSD (specific stem density) is the dry weight of a section of a plant's stem divided by the volume of the same section when still fresh expressed in mg/mm^3 (Cornelissen et al. 2003). Canopy height was measured as the distance between the highest photosynthetic tissue (leaf) and the base of the plant (Weiher et al. 1999).

Ecosystem property variables

Aboveground biomass (AGB) of the community was sampled on an area of 0.5 m^2 on each plot in March and August 2007, which is equal to beginning and peak biomass (De Leeuw et al. 1990). Samples were oven dried at 70°C for 72 hours and weighted. All values of biomass samples were doubled to refer to 1m^2 per plot. AGB refers to peak aboveground biomass. Aboveground net primary productivity (ANPP) is the difference between living biomass at the peak of the growing season and its beginning, divided by the period of time (here, five months). For this, standing biomass samples were sorted according to live and dead plant material, following 'method 3' of Scurlock et al. (2002).

Decomposition rates of community litter integrate the effects of litter quality, soil organisms and the environment on the carbon and nutrient cycle within the habitat. Preparation of litter bags followed Garnier et al. (2007), using a 1mm mesh. Litter bags were placed on the soil surface of the plots in which plant material was collected in autumn 2007 and were harvested after 6 and 12 months with four replicates per harvest, respectively. Extraneous plant material, animals and soil agglomerates were removed from the samples, and litter bags' content was oven dried at 60°C for 72 hours and weighted. Decomposition is the amount of decomposed material in percent. Species richness was evaluated as number of species found on each plot (Table 9-1).

Table 9-1: Environmental, trait, and ecosystem property variables used for this study. Groundwater level and salinity were aggregated by PCA-analysis; sample scores of first axis were named 'Groundwater'. Same was done for carbonate, phosphorous and soil sand content ('Nutrients'), and SLA, LDMC, SDMC and SSD (leaf and stem properties, 'LSP'). *: log-transformation of variables prior to path analysis.

Environmental variable	Abbreviation	Unit
Mean level of groundwater	Groundwater	cm
Mean salinity of groundwater		PSU
Inundation frequency	Inundation *	hours
Carbonate		t/ha
Phosphorus	Nutrients	kg/ha
Soil sand content		%
Trait variables	**Abbreviation**	**Unit**
Aboveground dry mass (dry weight of stems and leaves)	ADM	g
Belowground dry mass (dry weight of roots and rhizomes)	BDM *	g
Canopy height	Canopy	cm
Specific leaf area (SLA)		mm²/mg
Leaf dry matter content (LDMC)	Leaf and stem	mg/g
Stem dry matter content (SDMC)	properties, LSP	mg/g
Stem specific density (SSD)		mg/mm³
Ecosystem properties	**Abbreviation**	**Unit**
Aboveground biomass	AGB	g/m²
Aboveground primary productivity	ANPP	gC/m²/month
Decomposition after 6 months	Decomposition 6	%
Decomposition after 12 months	Decomposition 12	%
Number of species	Species richness	number

Statistical Analysis

Groundwater level

Groundwater level lacked information about tidal variation, as field measurements were restricted to low tide. A regression was conducted with paired values of the hourly data produced by data loggers at the nine plots and the biweekly data of the groundwater levels at all plots to fill this gap of knowledge. Mean values of both groundwater and logger data were used for linear regression analysis. The regression function was used to adjust values of mean groundwater level of all other plots to include information about high tide.

Community weighted mean

Our trait values were collected at the species level. To be able to predict ecosystem properties from traits it was necessary to scale up the traits to the community level, following Grime (1998). The mean value of each trait of every species was weighted by the abundance of the species on each plot and averaged by the total abundance of every species on each plot.

PCA of environmental parameters and plant trait variables

To reduce the variables in the models principal component analysis (PCA) was conducted with the following variables: mean level of groundwater and groundwater salinity were aggregated to 'groundwater'; soil sand content, plant available phosphorus and carbonate to 'nutrients'; and SLA, LDMC, SDMC and SSD to 'leaf and stem properties, LSP' (see Table 9-1). Analysis was carried out using the computer software R (The R Foundation for Statistical Computing 2008).

Path analysis and Structural Equation Modelling (SEM):

Our hypothetical model only included observed variables and is thus considered manifest (McCune and Grace 2002). Hence, we used path analysis to quantify the relationships between environmental conditions, plant functional traits and ecosystem properties. SEM was used to test overall model fitness (Grace and Pugesek 1998).

Path analysis partitions the correlation among variables and measures both direct and indirect effects on response variables (Grace 2006). The expected covariance structure of the hypothetical model is compared to the actual covariance matrix. Standardized coefficients describe the strength of the relationships. Relationships between variables are either unanalyzed correlations (indicated by a curved, double-headed arrow) or uni-directionally causal (indicated by a straight, single-headed arrow). Indirect pathways between variables involve intermediary variables. Direct pathways are the value of path coefficients, which are the standardized partial regression coefficients of the displayed arrow that directly connect two variables (Grace 2006).

SEM allows testing the hypothesis that the measured covariance structure adequately describes the expected covariance structure. This is done by means of maximum likelihood estimates which generate a test statistic that is distributed approximately as χ^2 (Backhaus et al. 2003). Good fit of the hypothesized model to the data will result in non-significant p-value. Goodness-of-fit-index (GFI) is a measure of the relative amount of variance and covariance that the model allows for. Another index of fit, the root mean square error of approximation (RMSEA) assesses closeness to fit. Good models have GFI > 0.9, p > 0.05, and RMSEA < 0.05-0.08 (good to fair model) (Backhaus et al. 2003). We used Amos 16.0.1 (Arbuckle 2007) to conduct path analysis and SEM for model evaluation.

9.3 Results

PCA of environmental parameters

Explained variance for 'groundwater', i.e. the 1st principal component for groundwater level and salinity was 0.81 for the first and 0.19 for the second axis. High PCA-values correspond to high groundwater level and salinity, low values to low groundwater level and low salt concentration. Explained proportion of variance for 'nutrients' and 'LSP' were lower with 0.54 and 0.29 (nutrients), and 0.91 and 0.04 (LSP) for the first two axes, respectively. Low PCA-values of nutrients refer to high sand content of the soil, high PCA-values to high amounts of phosphorus and carbonate. Low PCA-values for LSP indicate species with high SLA, whereas high values refer to high dry matter content of leaves and stem (high LDMC and SDMC) and high stem density. Sample scores of the first PCA-axis were used for further path analysis.

Trait expressions of the community and ecosystem properties

The amount of aboveground dry mass (ADM, community-trait) was almost similar in the lower and the upper marsh, BDM and canopy height were highest in the lower marsh (Figure 9-3) and decreased towards the pioneer zone and the upper marsh. PCA-scores of LSP showed that dry matter content of upper marsh species was highest in the upper marsh and lowest in the lower marsh.

Aboveground biomass (AGB, ecosystem property) was highest in the upper marsh and lowest in the pioneer zone. Productivity (ANPP) was lowest in the pioneer zone and highest in the upper marsh, however, several plots showed negative productivity values, which means that living biomass was higher in spring than in summer 2007. Decomposition rates after six months were lowest in the upper marsh, after 12 months almost all litter was decomposed in the three salt marsh zones. Species richness was lowest in the upper marsh and highest in the pioneer zone. The reason for the high species number in the pioneer zone is that although dominated by species like *Salicornia europaea*, this zone was also populated by lower marsh species, which led to an increase in total species richness.

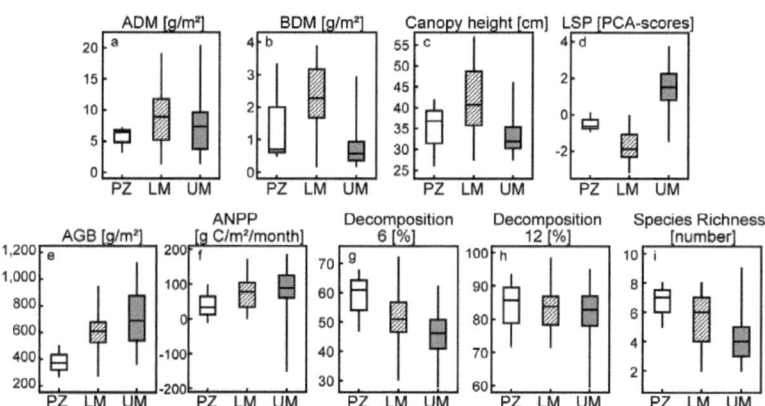

Figure 9-3: Boxplots of trait expressions of the community (first row) and ecosystem properties (second row) within the different salt marsh zones. Classification of zonation follows Pott (1995). PZ: Pioneer Zone, LM: Lower Marsh, UM: Upper Marsh.

Path analysis and Structural Equation Modelling (SEM)

Multivariate kurtosis was obtained in all models using Mardia's coefficient as indicator (Mardia 1970). Some direct paths proved non-significant and were deleted from the model (BMD → canopy (trait-trait relationships); nutrients ↔ inundation (correlations between environment); groundwater → canopy, LSP; inundation → canopy, BDM; nutrients → LSP, ADM, BDM (environment-trait relationships); ADM → ANPP, decomposition 6 and 12; BDM → decomposition 6 and 12; LSP → AGB, ANPP; canopy → species richness, AGB, decomposition 6 and 12 (trait-ecosystem properties relationships)).

These changes led to five stable models, which only included significant pathways at $p < 0.05$. The resulting models showed good consistency with the data and explained 44 % of the variation in species richness ($p = 0.38$, GFI: 0.96, RMSEA: 0.03), 16 % in AGB ($p = 0.79$, GFI: 0.97, RMSEA: 0.0), 21 % in ANPP ($p = 0.80$, GFI: 0.97, RMSEA: 0.0), 21 % in Decomposition 6 ($p = 0.15$, GFI: 0.94, RMSEA: 0.07), and 10 % in Decomposition 12 ($p = 0.74$, GFI: 0.97, RMSEA: 0.0).

All models consisted of the same trait and environmental variables and their relationships were identical in all models. Models differed only in the traits explaining the various ecosystem properties. The structural equation models revealed both direct and indirect effects between environmental conditions, traits and ecosystem properties. Direct effects of trait-trait relationships are visualized in Fig. 9-4a, between environment and traits in Fig 9-4b and between traits and ecosystem properties in Fig. 9-4c-g. Indirect effects occur if two variables are connected through paths to and from a third variable. Total effects are calculated from the combined direct and indirect effects. Standardised total, direct and indirect effects are listed in Table 9- 2.

Trait-trait linkages

Corresponding to our hypothesised model, LSP showed trade-offs with canopy height, above- and belowground dry mass, whereas ADM scaled positively with canopy height (Table 9-2). The relationship between BDM and ADM proofed positive, meaning that high mass of belowground organs provides the basis for aboveground growth.

The direct path from BDM to canopy height was deleted due to non significance, however a moderate indirect relationship is mediated through ADM. This allometric relationship is opposite to our expectations.

Response traits

The double-headed arrows between the environmental variables in Fig. 9-4b showed correlations between these variables. Nutrients were negatively correlated to groundwater level and salinity, indicating that areas of the lower marsh with high groundwater level and salinity are less rich in nutrients than the upper marsh areas. Groundwater level and salinity increased with inundation frequency, but the correlation was not as strong as expected.

According to our expectations groundwater and inundation exerted a negative effect on ADM and nutrients a positive effect on canopy height (see total effects in Table 9-2). Opposite to our initial model, BDM responded positively to groundwater and inundation. Also, contrary to what we expected, canopy height responded positively and LSP negatively to inundation frequency.

The paths from groundwater and nutrient availability to LSP were deleted from the final model, indicating that these environmental parameters are of less importance for the structural composition of the tissue than inundation frequency. Also, nutrients did not significantly influence ADM and BDM, meaning that groundwater and inundation frequency are the primary environmental constraints for these trait expressions in salt marsh plants.

Effect traits on ecosystem properties

Aboveground dry mass had a positive, belowground dry mass had a negative effect on AGB (Fig. 9-4c, Table 9-2). The effect of LSP was only weak. This ecosystem property is driven by both lower marsh species with high amounts of aboveground dry mass and upper marsh species producing small belowground dry mass (Fig. 9-3).

BDM and canopy height had greatest effects on ANPP (Fig. 9-4c), again, the effects of LSP were only weak. As with ADM, trait features of both upper and lower marsh exerted greatest effects on this ecosystem property.

Decomposition rates after 6 and 12 month were both determined by LSP, i.e. the higher the dry matter content, the slower the decomposition rates (Fig. 9-4e and f). LSP-values were greatest in the upper marsh, which means that this ecosystem property is most strongly affected by the traits of upper marsh plants.

Species richness was mostly determined by the effects of LSP and above- and belowground dry mass (Fig. 9-4g). In areas in which species produced tissues with high dry matter content and high aboveground dry mass (upper marsh, Figure 9-3), species richness was lower than in areas in which species showed high belowground dry mass (lower marsh). These relationships were opposite to our expectations.

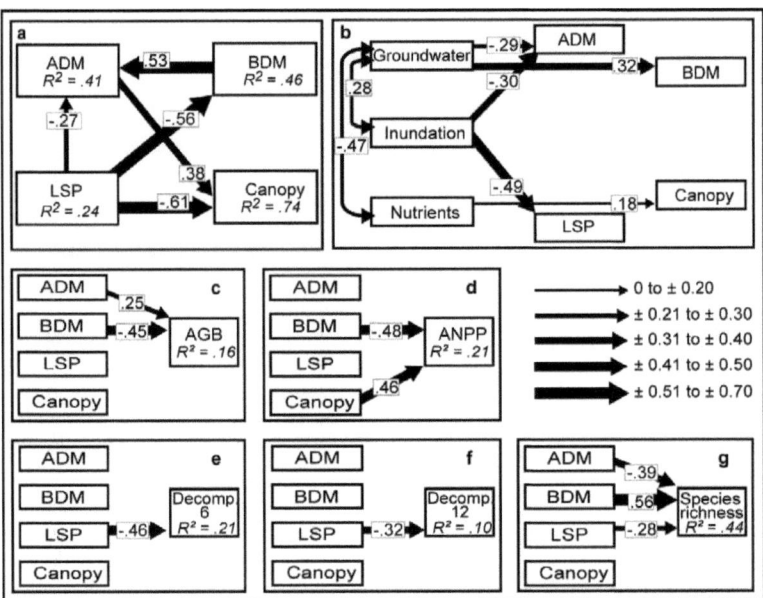

Figure 9-4: Final models derived from initial models in Figure 9-1. Names and abbreviations of observed variables follow Table 9-1. Path coefficients between variables are standardized partial regression coefficients of direct effects, for total and indirect effects see Table 9-2. Arrow widths are proportional to the standardized path coefficient. Variances explained by the model (R^2) are given under the variable names. All pathways are significant at p < 0.05. a: trait-trait relationships, b: environment–trait relationships; trait relationships to AGB (c), ANPP (d), Decomposition 6 (e) Decomposition 12 (f) and to species richness (g).

Table 9-2: Standardized total, direct and indirect effects of (i) environmental parameters on functional traits, (ii) of trait-trait relationships, and (iii) of environmental parameters and functional traits on ecosystem properties. Total effects were calculated by adding direct and indirect effects. All effects were significant at p < 0.05. For names and abbreviations see Table 9-1.

Total, direct and indirect effects of environmental parameters on functional traits and total, direct and indirect effects of trait-trait relationships

	Groundwater			Inundation			Nutrients			ADM			BDM			LSP			Canopy height		
	Tot.	Dir.	Ind.	Tot.	Dir.	Ind.	Tot.	Dir.	Ind.	Tot.	Dir.	Ind.	Tot.	Dir.	Ind.	Tot.	Dir.	Ind.	Tot.	Dir.	Ind.
ADM	-.12	-.29	.17	-.02	-.30	.28															
BDM	.32	.32		.28		.28										-.57	-.27	-.30			
LSP				-.49	-.49		.18	.18					.53	.53							
Canopy	-.05		-.05	.29		.29				.38	.38		.20		.20	-.83	-.61	-.22			

Total, direct and indirect effects of environmental parameters and functional traits on ecosystem properties

	Groundwater			Inundation			Nutrients			ADM			BDM			LSP			Canopy height		
	Tot.	Dir.	Ind.	Tot.	Dir.	Ind.	Tot.	Dir.	Ind.	Tot.	Dir.	Ind.	Tot.	Dir.	Ind.	Tot.	Dir.	Ind.	Tot.	Dir.	Ind.
AGB	-.17		-.17	-.13		-.13				.25	.25		-.32	-.45	.13	-.11		-.11			
ANPP	-.17		-.17				.08		.08	.18		.18	-.39	-.48	.09	-.11		-.11	.46	.46	
Dec. 6				.23		.23										-.46	-.46				
Dec. 12				.16		.16										-.32	-.32				
Sp. Richn.	.23		.23	.30		.30				-.39	-.39		.35	.56	-.21	-.38	-.28	-.10			

9.4 Discussion

Finding key traits relevant for ecosystem properties is a major task of functional ecology. In our study we found that (i) trait responses to the environment and their effects on ecosystem properties are in part opposite to expectations derived from common scientific knowledge, and (ii) different traits are affecting different ecosystem properties. The latter highlights the relevance of phenotypic trait diversity to maintain different ecosystem properties.

Relationships among traits and responses to environmental conditions

ADM responded positively to BDM, which means that a high mass of belowground organs provides the basis for aboveground biomass production (Enquist and Niklas 2002). While some species with high allocation to BDM showed high aboveground biomass and high canopy height, others showed the same allocation to BDM, however were less tall in stature. BDM was high in the lower marsh (positive response to inundation frequency and groundwater) which supports the findings of Bakker et al. (2003). They state that plants growing in the lower marsh show higher root mass than those of the upper marsh and attribute this to a strategy to anchor the plants in the sometimes unstable sediment and provide resistance to water currents and wave energy during inundations. We assume two different strategies of plants to colonize the lower marsh: for some species allocation to belowground biomass is a prerequisite for gaining high canopy height and with that high competitive ability, while others, in particular rosette plants, show high BDM and small vertical growth.

Our final model revealed negative correlations between groundwater level and salinity and nutrient availability, meaning that soils of elevated sites with low groundwater level showed higher nutrient content. This is in accordance with the findings of Olff et al. (1997) who found an increase of nitrogen of later successional stages where more clay had accumulated. Furthermore, groundwater level and salinity were positively correlated with inundation frequency, although correlation was only moderate.

The positive effect of inundation frequency on canopy height was contrary to our assumptions, as well as the negative response of LSP. We interpret the fact that lower marsh species showed high SLA and low LDMC, SDMC and SSD to be an adaptation to salt stress (dilution of cell sap and succulent growth). Leaf trait-environmental relationships of salt marsh species are inverse to those of other terrestrial plants, meaning that the economics spectrum derived from leaf traits of other terrestrial plants does not hold in salt marshes (Wright et al. 2004, Minden et al. unpubl.). Leaf traits of salt marsh plants are primarily constrained by groundwater and inundation; our study could not detect a significant relationship between nutrient availability and leaf traits.

Traits effecting ecosystem properties

AGB

We expected a strong effect of ADM on AGB, as community biomass should be the sum of the aboveground dry mass of the species composing the community. As BDM allometrically scales with ADM there should be a positive effect of BDM on AGB as well. However, the effect of ADM on AGB was weaker than that of BDM, which was negative rather than positive. Indirect effects of inundation on AGB were also negative meaning that AGB was higher on the upper marsh (see also Fig. 9-3). This was also found by Bakker et al. (2003, Minden et al. unpubl.).

However, as mentioned above, physical stress at the lower marsh enforces stronger allocation to roots and rhizomes whereas the more benign conditions of the upper marsh allow increasing shoot biomass. Note that these results depended on the high abundance of a few species only, such as *Limonium vulgare* or *Atriplex portulacoides* with their large rhizomes on the lower marsh. Thus, on the relatively small scale of a salt marsh, adaptations to environmental conditions override a positive relationship between BDM and ADM on AGB which could be expected based on allometrical relationships found in large datasets that however do not take environmental conditions into account (e.g. Enquist and Niklas 2002).

ANPP

Trait expressions of both the upper (low values of BDM) and the lower marsh plants (canopy height, weak negative effect of LSP Fig. 9-3, Table 9-2) explained ANPP. The weak effect of LSP to ANPP is in agreement with Pontes et al. (2007), who found SLA and ANPP poorly correlated. Also AGB only exerted weak effects on ANPP. The strongest effects on productivity were shown by canopy height (positive) and BDM (negative). Species of the lower marsh either produce a high canopy or allocate biomass to belowground organs or both. An elevated canopy has significant consequences on the intensity and quality of radiation which might affect successful growth and reproduction (Grime 2002). For anchoring the plants in the unstable sediment of the lower marsh, especially when the aboveground plant body is tall, species have to produce accordingly higher belowground biomass (Bakker et al. 2003). This explains the positive total effect of BDM on canopy height. However, other species of the lower marsh produce high BDM, but are only small in stature, which leads to a negative direct effect of BDM on ANPP. Also, less allocation of upper marsh species in belowground organs drives this relationship.

Species of the lower salt marsh show two different strategies, one of high allocation to BDM and high canopy height and ANPP and another one of nutrient conservation and low productivity. In the upper marsh, species allocate less biomass to belowground organs, but produce more aboveground mass.

Decomposition after 6 and 12 months

Our final models clearly supported our expectations that the higher the LSP of salt marsh plants the lower the litter decomposition rates. These results were also found in the Mediterranean region of France by Kazakou et al. (2006).

Inundation frequency effected decomposition positively, which means that litter of the pioneer zone and lower marsh decomposed more rapidly than the litter of the upper marsh, due to higher SLA and lower LDMC of the species of these zones. This relationship has been documented for other ecosystems as well (Garnier et al. 2004).

Species richness

In our hypothesised model we assumed LSP to positively explain species richness, because the association of the aggregated traits with slow RGR should lead to low competitive exclusion and with that to high species richness. Species of the upper salt marsh showed highest values of LSP (Fig. 9-3), but these sites were mostly dominated by *Elymus athericus* and only few others (Pott 1995), whereas species number is highest in the lower parts of the salt marsh. Aboveground dry mass also showed negative effects on species richness, whereas belowground dry mass affected species richness positively. The latter is highest in the pioneer zone and the lower marsh, which both show highest species richness.

Grace and Pugesek (1997) evaluated the effects of abiotic conditions on species richness via path analysis, and found that areas with high light penetration ($< 75\%$, pioneer zone) and very low light penetration had lowest species richness. They explained the drop in species richness at low light with exclusion due to shading. We support their explanation insofar as we found that high LSP decreases decomposition in the upper marsh (Fig. 9-4 e, f), which leads to an accumulation of dead biomass, which again triggers low light penetration. However, we cannot support their findings for the pioneer zone.

Conclusions

Our study demonstrated that various ecosystem properties are dependent on different trait attributes of salt marsh species. Keystone response traits were (i) BDM responding to groundwater level and salinity positively and (ii) LSP responding to inundation frequency negatively. Leaf and stem properties (LSP) represented the most relevant combination of traits affecting the ecosystem properties in this study, however it is not exclusively these components of the plant economics spectrum (Freschet et al. 2010) which can be held responsible for explaining AGB, ANPP, decomposition rates and species richness in salt marshes. More likely the interaction of different traits contributes to the functioning of this ecosystem. Another important result of our study is the response of traits to direct environmental factors such as salinity and soil resources which separate the salt marsh into the pioneer zone, lower and the higher marsh.

Other studies showed the relevance of species diversity for maintaining multiple ecosystem functions (e.g. Hector and Bagchi 2007). This study verifies the importance of functional diversity for ecosystem multifunctionality. We could distinguish several functional traits responsible for different ecosystem properties and could attribute them to upper and lower marsh areas. It should be noted that the species pool of these areas is very small. Loss of certain species has the potential to change the relationships between environment and vegetation-based ecosystem properties considerably and might thus have profound effects on the functioning of the whole ecosystem and those adjacent and depending on it, like e.g. the Wadden Sea system.

Acknowledgements:

We thank the administration of the National park 'Niedersächsisches Wattenmeer'. This study was conducted as part of the TREIBSEL project and was supported by the 'II. Oldenburgischer Deichband' and the 'Wasserverbandstag e.V.' (NWS 10/05).

10 Synthesis

10.1 General remarks

Protection of habitats and preservation of biodiversity as a basis for ecosystem functioning is a major goal of conservation biology, particularly against the background of increasing human-induced habitat destruction. However, conservation of species richness *per se* in order to maintain ecosystem functioning without understanding the underlying principles of the relationships of environmental constraints, expressions of plant traits as a response to the environment and their effects on ecosystem properties is a forlorn hope, as it ignores that ecosystem functioning is an outcome of the combination of the mentioned processes rather than just a construction of biodiversity.

This thesis uses the plant functional trait approach to determine the influence of environmental conditions on plant trait expressions and evaluates their effects on ecosystem properties ultimately aiming at drawing conclusions of the role of biodiversity on ecosystem functioning in salt marshes.

The following paragraphs synthesize the outcomes of the analyses which were presented in the previous chapters. Rather than summarizing each analysis separately (chapter 6 to 9), the next passages were arranged so as to accentuate the role of environmental conditions in salt marshes, the response of salt marsh plants determined from their trait expressions, the effects of plant traits on ecosystem properties, and the role of biodiversity for the maintenance of ecosystem properties. Furthermore, possible results of statistical analyses using different parameter-combinations are introduced. The last two paragraphs address possible improvements in methodology as well as need and suggestions for further research, respectively.

10.2 Environmental constraints and species distribution

It was stated earlier in the text that soil potassium and phosphorus contents decrease towards the upper marsh, as found in two salt marshes in the UK (Ranwell 1964, Gray and Bunce 1972). Also carbonate was expected to be highest in the lower areas of the marsh (see Adam 1990). However, these finding cannot be supported for the mainland marshes of this thesis, for which no significant differences were found between the pioneer zone, lower and upper marsh for potassium and phosphorus and no difference between pioneer zone and upper marsh for carbonate (chapter 5.4, Fig. 5-4). Considering the marsh on the island of Mellum, a distinct decrease of nutrients towards the upper marsh like those found in the UK cannot be seen either, however the patterns seem more suggestive than those of the mainland.

This means that within mainland and island marshes there is no distinct nutrient-related gradient. However, the data clearly demonstrate that the soils of mainland marshes are richer in nutrients than those of islands, which contributed to a strong nutrient gradient spanning from mainland to island salt marshes in chapter 6.

Environmental parameters related to the water regime showed similar patterns both on the mainland and on the island. Inundation frequency and groundwater level and salinity decreased towards the upper marsh and showed highest values in the pioneer zone.

Results of RLQ-analysis (chapter 6) revealed that species performance in salt marshes is influenced mainly by those two gradients and utilization by land use (mowing, cattle grazing) plays only a minor role. However, it seems that whereas the nutrient-related gradient separates mainland from island marshes, the water-related gradient can be more distinctively identified within both mainland and island marshes, i.e. from pioneer zone to upper marsh, which might be the reason why those two gradients were uncorrelated as result of the RLQ-analysis.

It was possible to detect patterns of species distribution which follow these gradients (clustering of species scores of first two CCA-axes, chapter 7). Across marshes (from mainland to island marshes) species are separated by the nutrient-related gradient, i.e. nutrient rich (mainland) to nutrient poor (island) sites. Within a marsh, species are separated by the water-related gradient, i.e. frequently inundated sites (with high groundwater level and salinity) and infrequently inundated sites (with low groundwater level and salinity).

10.3 Trait expressions of salt marsh plants as response to environmental conditions

Each analysis (chapter 6 to 9) was conducted addressing different questions and the traits chosen for each analysis differed in respect to those questions. In order to summarize the most important response traits of salt marsh plants a fourth corner-analysis was conducted using the most significant response traits of each of the separate analysis. This was done for mainland and island marshes separately (Appendices 7 and 8) and for the 'whole dataset' in which mainland and island data were brought together (Appendix 9).

The most striking difference between trait expressions of mainland and island plants in reference to environmental conditions is that traits of mainland species are less significantly correlated with the nutrient-related environmental parameters (plant available phosphorus, potassium, carbonate and sand content of the soil). Plant traits of island species seem more constrained by nutrient availability than those of the mainland marshes (particularly LDMC, SDMC, C:N ratio of stem and diaspores and C:P ratio of diaspores), which is actually not very surprising given the fact that mainland marshes are richer in nutrients than island marshes, and also the distribution of nutrients along the elevation gradient is more balanced on the mainland (see chapter above).

Trait expressions in reference to water-related environmental parameters (inundation frequency, groundwater level and salinity) show similar patterns in mainland and island salt marsh plants, despite those reflecting the stoichiometry of the different plant organs. Here, C:N and C:P ratios of mainland plants are more significantly correlated with the water-related parameters, which might also be associated with increased nutrient availability in those marshes. Nevertheless, it seems plant trait expressions of mainland and island salt marsh plants show similar patterns in respect to water-related environmental conditions and differ in respect to nutrient-related conditions.

Taking both mainland and island species into account (Appendix 9) distinct patterns of trait expressions with reference to environmental parameters can be detected, which repeat, support and summarize the findings of the analyses accomplished before (chapter 6 to 9). A set of morphology-based plant traits both respond most strongly to nutrient- and water-related environmental parameters; these are (nutrient-related): canopy height, reproductive effort (RE), stem, leaf and root mass fraction (SMF, LMF and RMF) and (water-related): exclusion of salt ions[1] (Excl.), morphological adaptations to osmotic stress (MAOS) and leaf dry matter content (LDMC). These results were also found in the chapters 6, 8 and 9. Elemental-based traits (C:N, C:P and N:P ratios of plant organs) either respond most strongly to the water-related (C:N ratio) or to nutrient-related environmental parameters (C:P and N:P ratios), which was already observed in chapter 7.

10.4 Possible soft traits as surrogates for hard traits

So-called 'soft' traits are traits that are easily accessible and relatively undemanding considering investment of time and money (Hodgson et al. 1999, Weiher et al. 1999). They can be used as surrogates for so-called 'hard' traits, which reflect the actual function of interest, but are difficult to obtain (Violle et al. 2007). Plant traits of this thesis were collected independently from the soft/hard trait classification. Nevertheless, some traits were relatively easy to measure (morphology-based traits) and might be used as surrogates to reflect the functional characteristic of traits that were of more laborious work (elemental-based traits) for future projects on salt marshes.

As no distinct analysis was conducted which referred to the 'soft/hard trait surrogate-approach', it can only be speculated about possible outcomes. However, based on the results of the previous analyses (especially chapter 6 and 7) it can be expected that LDMC and MAOS might serve as

---

[1] Exclusion of salt ions could also be referred to as physiology-related trait, but is based on the existence of glands and bladders, which are morphological features.

surrogates for carbon-rich, nutrient poor structures (i.e. those with high C:N ratio) and exclusion of salt ions (Excl.) would be positioned at the opposite site of the gradient e.g. in an ordination diagram of a RLQ-analysis (Fig. 10-1). However, it would be expected that LDMC and Excl. more strongly reflect the carbon content than the nutrient content of the tissues.

As C:P and N:P ratios of plant organs proofed to respond more to the nutrient status of the environment, they are expected to show high correlations to morphology-based traits which are orientated along the nutrient gradient (positive correlation to leaf and root mass fractions (LMF and RMF) and negative to canopy height, reproductive effort (RE) and stem mass fractions (SMF)). The internal regulation of stem C:N:P ratios proofed to be more heterostatic than those of the other plant organs, so that either morphology-based traits responding to the water- or to the nutrient-related environmental parameters might serve as surrogates.

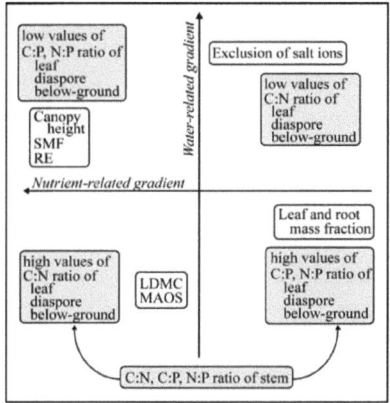

Figure 10-1: Possible arrangement of morphology-based ('soft') traits and elemental-based ('hard') traits along the nutrient-related and water-related gradients. Position of stem C:N, C:P and N:P ratios might be either on the left or the right side of the diagram, as indicated by two arrows.

Taking the quantitative surrogates (LDMC, canopy height, stem, leaf, root mass fraction (SMF, LMF, RMF) and reproductive effort (RE)) into account, what can one can expect when dividing the dataset into four subgroups (nutrient rich, infrequently inundated sites vs. nutrient rich, frequently inundated sites vs. nutrient poor, infrequently inundated sites vs. nutrient poor, frequently inundated sites) and rerunning the bivariate-line fitting technique introduced in chapter 7?

When focusing on differences regarding the influence of nutrient availability under a similar water regime (i.e. nutrient rich, infrequently inundated sites vs. nutrient poor, infrequently inundated sites and nutrient rich, frequently inundated sites vs. nutrient poor, frequently inundated sites) and using canopy height (or SMF or RE) for the y-axis and LMF (or RMF) for the abscissa one should get a) an insignificant result for nutrient rich, infrequently inundated vs. nutrient poor infrequently inundated sites and b) a shift along the axis for nutrient rich, frequently inundated vs. nutrient poor frequently inundated sites, see Fig. 10-2a. However, these considerations do not take into account the effects of outliers and residuals, which also influence the significance of the results.

Why an insignificant result for infrequently inundated sites differing in their nutrient status? In chapter 6 species were clustered according to their trait expressions (using soft traits) in reference to their appearance and the environmental conditions on the survey plots. Species from the infrequently inundated sites, both under nutrient poor and rich conditions formed one single cluster, because they share similar trait expressions (Fig. 6-4, cluster C). If one separates this cluster according nutrient rich and poor site conditions, expressions of soft traits are still similar, which would result in two clouds arranged too close to each other to show significant differences (Fig. 10-2a, white clouds of points).

Why a shift along the axis for frequently inundated sites differing in their nutrient status? Using trait expressions of species from clusters A and B in chapter 6 (frequently inundated, nutrient rich sites) and those from cluster E (frequently inundated, nutrient poor sites) one recognizes high canopy height, high values of SMF and RE but low LMF and RMF for clusters A and B. Opposite to that, species from cluster E show a lower stature and low values of SMF and RE, but high values for LMF and RMF, which would position the cloud of points close to the x-axis with low values on the y-axis (Fig. 10-2a, grey clouds of points).

Concerning possible trait surrogates this means that the soft traits canopy height, SMF, RE, LMF and RMF can be used as surrogates for hard traits reflecting species from the frequently inundated sites (lower marsh) differing in the nutrient status of their environment, but not for those of the infrequently inundated sites (upper marsh).

The next focus lies on differences in regard to the influence of the water regime under similar nutrient conditions (i.e. infrequently inundated, nutrient rich, vs. frequently inundated, nutrient rich sites and infrequently inundated, nutrient poor vs. frequently inundated, nutrient poor sites). There is unfortunately no quantitative parameter which could be used for representing species from the frequently inundated sites, but as RMF shows also a high correlation to the water-related parameters (Appendix 9), this trait seems appropriate. Values of RMF are plotted against LDMC-values separated into the four clusters.

Both the RMF- and LDMC-values for species from the frequently inundated, nutrient rich sites (clusters A and B, chapter 6) are lower than those of the infrequently inundated, nutrient rich sites (cluster C), which should result in a shift in elevation and a shift along the axis for those clusters as presented in Fig. 10-2b (white clouds of points). The same should result when plotting RMF- and LDMC-values from species from nutrient poor, infrequently inundated sites against those from nutrient poor, frequently inundated sites, however, here species from the lower parts of the marsh should show higher values of RMF and lower of LDMC than those of the upper marsh.

Based on this it is assumed that the soft traits LDMC and RMF can serve as surrogates for expressing hard traits of species from infrequently and frequently inundated sites, respectively.

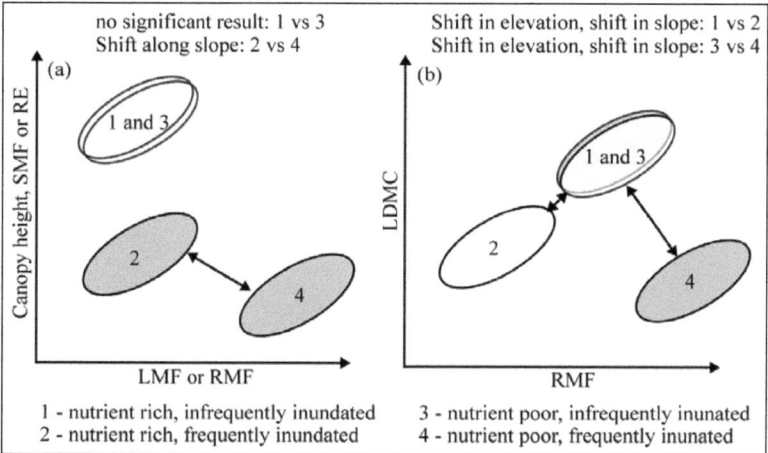

Figure 10-2: Possible alignments of clouds of points using 'soft' traits as surrogates for 'hard' traits and orientating on the outcomes of chapter 6 and 7.

To sum up it is assumed that the soft trait LDMC can be used as a surrogate for plant traits of carbon-rich, upper marsh species and RMF as surrogate for traits of carbon-poor, lower marsh species. Canopy height, RE, SMF and LMF could be seen as surrogates for trait expressions of nutrient rich, carbon poor plant traits.

10.5 Trait-trait relationships

Relationships between traits were already discussed in chapters 6 to 9, and especially chapters 8 and 9 put special emphasis on this topic. Summing up the various outcomes of the analyses, it is possible to show plant trait expressions which are closely related to each other, while between others there is no or a negative relationship.

As already mentioned above trait expressions of canopy height, SMF and RE, as well as LDMC and MAOS, and RMF and LMF were similar between certain sets of species. Another analysis (chapter 8) revealed a close relationship between C:N ratio of the whole plant and MAOS, SLA and dry mass of belowground organs, and dry weight of leaves, stems, diaspores and belowground organs. Opposite to that, C:N ratio of the whole plant and MAOS were found to be negatively related to SLA, and MAOS negatively to dry weight of leaves. Finally, chapter 9 describes close relationships between dry weight of aboveground plant organs to those of belowground organs, whereas a negative relationship could be detected between the latter two and 'leaf and stem properties' (aggregated by SLA, LDMC, SDMC and SSD).

Although not statistically verified yet, there might be various significant relationships between plant traits (Fig. 10-3, note that the arrows do not imply casual relationships). As already seen in chapter 6 canopy height, RE and SMF were positioned at the opposite sites from LMF and RMF along the nutrient-related gradient in the RLQ-space, which makes a significant negative relationship between these traits very likely. As allocation implies trade-offs (see chapter 3.5, Weiner 2004) a plant that invests heavily in its stem (which results in high stem mass) could be expected to have a high stem mass fraction, hence a negative relationship between dry weight of stem and SMF (this also counts for the relationships between dry weight of belowground mass and RMF, and dry weight of leaves and LMF). As aboveground dry mass (ADM) consist of the dry weight of stems, leaves, diaspores and belowground organs, a positive relationship between those traits can be expected. Results of other studies show that fast-growing species often show high SLA-values (Lavorel et al. 2007), so that one could expect SLA to be positively related to aboveground dry mass (ADM). However, the analyses in this thesis (chapter 6) indicate that salt marsh plants show different trait expressions in respect to SLA than other terrestrial plants, so that no positive relationship of SLA and ADM can be expected at the outset. As was already shown in the previous chapters, plant species with carbon rich tissue also show high LDMC-values and colonize the infrequently inundated sites of the upper marsh. This makes a positive relationship between LDMC and C:N ratio of the whole plant, as well as between C:N-ratio of the whole plant and LSP very likely.

The grey circles in Fig. 10-3 indicate the element-related hard traits of this thesis, for which the soft traits LDMC and MAOS, as well as LMF and RMF might be used as surrogates. The hard traits would accordingly show similar relationships to other traits than their soft trait surrogate.

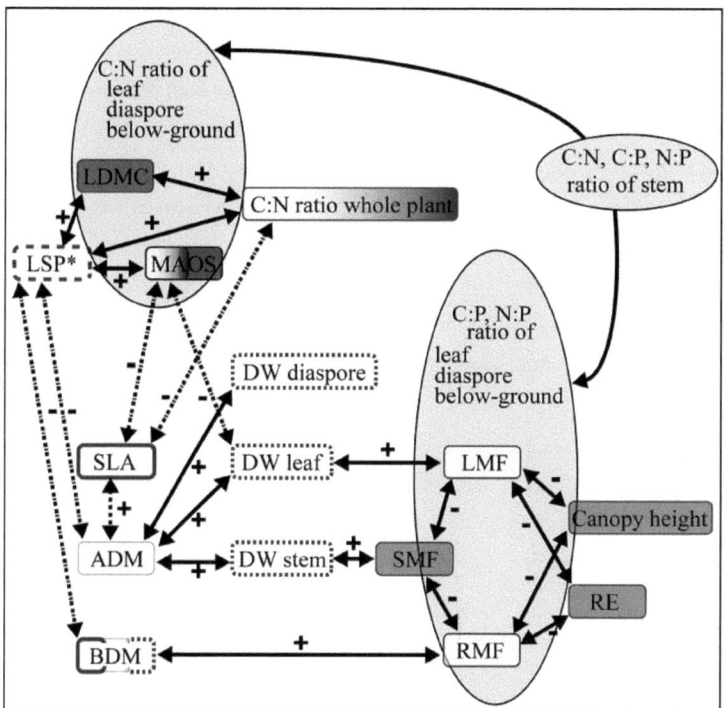

Figure 10-3: Overview of trait-trait relationships. Traits sharing same shading were found to be positively related in previous analyses, negative relationships found in chapters 6 to 9 are indicated by arrows (dash-dot line), see also text above. Black arrows indicate possible positive ('+') and negative ('-') relationships between trait expressions which were not tested yet. The dashed arrow indicates possible relationships, but of doubt. LSP*: aggregated from LDMC, SDMC, SSD and SLA, high LSP-values indicate high LDMC-, SDMC- and SSD-values. Grey circles indicate the position of hard traits with similar relationships to other soft traits.

10.6 Plant traits effecting ecosystem functioning and the role of biodiversity

One purpose of this thesis is the evaluation of plant traits effecting ecosystem functioning (chapter 8 and 9). It was possible to show that various plant traits are responsible for different functions and in that course the role of biodiversity for the maintenance of ecosystem functioning was highlighted. However, only certain ecosystem functions could be addressed and the effects of plant traits on them statistically verified. In the following, plant traits which supposedly most strongly affect various ecosystem functions are listed and presumptions on the influence of biodiversity are given.

Carbon storage:

As already shown in chapter 9, plant traits like LDMC, SDMC and SSD (aggregated in LSP) explained decomposition rates most strongly; that means the higher the e.g. LDMC-values of plant tissue, the lower the decomposition rates of plant litter. These trait features are most strongly expressed by higher marsh species like *Elymus athericus* and *Festuca rubra* (both mainland and island). It can further be assumed that carbon storage over longer periods of time is highest in areas with low decomposition rates. It is therefore expected that carbon storage would be most strongly affected by the same traits that are responsible for low decomposition rates and furthermore by high tissue C:N and C:P ratios. Studies with special focus on carbon storage (for example in relation to climate change) could use those plant traits for e.g. their models. In regard to biodiversity, higher marsh species would be crucial for long-term carbon storage in salt marshes.

Food webs:

As already pointed out (chapter 4.4) salt marshes are inhabited by large numbers of e.g. invertebrates and provide feeding ground for many bird species (Heydemann 1981, Blew et al. 2005). But these organisms are only single chains in the whole food web of an ecosystem, which is actually based on autotrophs providing the basis by fixing carbon, taking up nutrients and finally releasing back the elements into the system by decomposition or being directly consumed by herbivores.

Carbon-poor, nutrient-rich tissue is more palatable for herbivores than e.g. lignin rich plant organs (De Deyn et al. 2008). It can therefore be expected that plant organs with high nutrient content are more attractive and also provide a more important basis for salt marsh food webs. Based on the findings of the analyses and the assumptions in chapters 10.3 to 10.5, low tissue C:N and C:P ratios as well as high values of SMF and RE are expected to affect palatability of plant tissue most strongly and with that have high influence on the food web of the whole marsh. As these traits are most strongly exhibited by lower marsh species, loss of species in this area might have profound effects on the food web of the whole ecosystem.

Coastal protection:

Wave attenuation is a crucial ecosystem service of salt marshes in regard to coastal protection. Bouma et al. (2005) found that in species from the intertidal zone (*Spartina anglica* and *Zostera noltii*) shoot stiffness determines the capacity to reduce hydrodynamic energy. However, it can also be expected that especially species from the upper marsh which exhibit high SDMC- and LDMC-values might also be important in reducing wave energy. Also, high tissue C:N-content should be strongly related to tissue stiffness.

Summarizing it can be concluded that the conservation of biodiversity and with that the functional characteristics of salt marsh plants is not only crucial for maintaining the ecosystem properties that were already discussed in chapters 8 and 9, but also for other ecosystem functions, for with statistical verification is pending.

10.7 Improvements on methodology

Measurement of sedimentation and erosion:

Evaluation of sedimentation and erosion rates was conducted by so-called Sedimentation/Erosion Bars (SEB's, chapter 5.4). Unfortunately, in the grazed areas the PVC-poles were used by the cattle to scratch themselves at and some were even damaged so that they could not be used anymore. Also, trampling of cows and increased destruction of vegetation around the poles influenced the resulting data. In the mown areas, the poles had to be fenced to prevent damage to the mowing-machines and ultimately had to be removed when the field survey was completed. Fencing might have also influenced sedimentation rates as the vegetation was allowed to grow in the otherwise cut down vegetation. There was no damage to the SEB's in the barren sites. Use of a lighter measuring stick might be of advantage as the one currently in use sinks into the ground too easily especially when the soil is soaked. As an alternative to the measuring stick, the use of a laser pointer which measures distances is suggested.

In grazed and mown areas, the so-called 'plate technique' could be used (Brown 1998). Aluminum or PVC plates are buried into the ground, and after an initial settlement phase the distance between surface and plate is measured and, after repeated measurements the sedimentation and erosion rates can be calculated. One advantage of this technique is that it is not visible in the landscape and that cattle is not so much attracted by it and mowing with machines is not influenced. One disadvantage is that it might be hard to find the places the plates were buried, especially in dense vegetation.

Measurements of carbon, nitrogen and phosphorus contents of plant organs:

Chemical analysis for determination of C:N content demands 2-3 mg plant material and for phosphorus analysis 7-8 mg plant material is needed. Available material met those conditions in most cases, however, dry weight of organs of some plant species was so low (e.g. leaves of *Festuca rubra* or diaspores of *Salicornia europaea*) that the analyses could not be conducted for each individual. Pooling of material of several individuals of one species in order to gain enough material for chemical analysis might result in inappropriate data for some statistical analyses (e.g. bivariate line-fitting, chapter 9). Therefore, alternative chemical analyses that require less material would be of advantage, however, no suggestions can be given at this point.

Measurement of plant available ammonium and nitrate:

In this thesis synthetic ion-exchanger were used to evaluate the amount of plant available ammonium and nitrate in the soil (Skogley and Dobermann 1996). Unfortunately, this technique proofed inappropriate for salt marshes, as Na^+-ions occupied the docking locations of ammonium and Cl^--ions those reserved for nitrate, so that the data could not be used for further statistical analyses. Therefore alternative methods for measuring ammonium and nitrate in salt marsh soils should be sought.

10.8 Recommendations for further analysis and research

Variability in element:element ratios:

Plant material from which the C:N:P ratios of the different organs was determined (chapter 7) was harvested at the end of the growing season, at which the seeds of the plants were ripe but not yet shed. The data hence reflect a single point in time and do not include variability due to seasonal changes. Some studies indicate that there is seasonal variability, for example increase in foliage N:P and N:K ratios during the growing season (Aronsson and Elowson 1980, Chidumayo 1994). However, others do observe variability but without any clear directional trend (Grigal et al. 1976, Andersson 1997) so that no conclusions can be drawn on possible reasons for stoichiometric variability in element:element ratios (Ågren 2008).

C:N ratios for plant organs in the vegetative stage were measured as part of this thesis, however, a comparison with C:N ratios of the generative stage regarding seasonal variation is still lacking.

The analysis conducted in chapter 7 concentrated on interspecific variation and response to differences in nutrient status and inundation frequency and groundwater level and salinity. However, in reference to salt marshes, light could be thought of as another factor which might influence the elemental composition plant organs, which might not play such an important role in other terrestrial habitats. During times of inundation, the available light is not only decreased, but sometimes cut off altogether, which might alter the photoperiod and by that the carbon-fixation (Adam 1990). It can be expected that light availability is closely related to inundation frequency and hence plants of the lower marsh areas are exposed to lower light conditions. However, this effect could also be diluted by decreased light conditions in the upper marsh due to an increased litter layer near the surface.

Another question which has not been addressed yet is intraspecific variation in element:element ratios of plant organs between individuals of one species. As a result of chapter 7 it could be shown that the nutrient and water status of an area influences the elemental composition of salt marsh plant organs and that C:N:P ratios of leaves, diaspores and belowground organs are more homeostatic than those of stems. However, a question not addressed so far is if there are differences within a species along the elevation gradient which are also constrained by environmental conditions.

It has been shown for a single species (*Festuca rubra*) that grazing pressure of geese increased along the elevation gradient, whereas the nitrogen content of leaves decreased with height (Stahl 2002). One could expect the same patterns for other marsh species.

Effects of soil compaction on decomposition rates:
The results of chapter 9 indicate no influence of soil sand content, phosphorus and potassium on the decomposition rates of plant litter. However, there is evidence that suggests an influence of soil compaction on mineralization rates in salt marshes (Pieter Heijning, personal communication). Bulk density was measured as part of this thesis, so a statistical verification of the effects of soil compactions on decomposition of salt marsh plant litter would be interesting.

Effects of utilization on plant traits:
Utilization (biomass removal and trampling by cattle grazing and mowing) only showed weak relationships to the traits expressed by salt marsh plants (chapter 6). This is because the stocking with cattle was relatively low in the studied areas (animal units 0.5 to 1.5 animals per hectare) and mowing was carried out only once per year, so that the disturbance by utilization was too weak to show any significant expressions in salt marsh plants. Increasing the stocking rates and frequency of mowing (or mimicking them by clipping and artificial trampling) could result in significant responses of plant traits, which could contribute to the knowledge about the effects of land-use on trait expressions in this habitat.

Mainland vs island species:
In this thesis the dataset was separated into mainland and island species because both environmental conditions and plant trait expressions differed greatly between them. However, genetic proof that island and mainland species belong to different populations has yet to be given. It would also be interesting to include plant species of other Wadden Sea islands to evaluate if they belong to the same or to different populations.

Threshold element ratio:
As already pointed out in chapter 3.4.1 the threshold element ratio (ratio at which a nutrient is considered limiting for plant growth, Frost et al. 2005) can be quite variable insofar as different species respond to nutrient additions in different ways (Sterner and Elser 2002). Vitousek and Howarth (1991) therefore suggest to conduct replicated nutrient addition experiments to define the threshold that applies to an ecosystem. Concerning salt marshes it could be expected that mainland species respond differently to nutrient addition that island species, thus conducting experiments on this topic could produce results valuable for understanding the responses of this ecosystem to different levels of nutrient input.

Focus on belowground biomass production:
Most of the biomass produced in salt marshes is found belowground (Groenendijk and Vink-Lievaart 1987, Bakker et al. 1993) and belowground competition is increased as response to low soil nitrogen-availability (Berendse and Möller 2009). As measuring belowground biomass is very time consuming, it was not possible to include it in the time frame of this thesis. Nevertheless, evaluating the effects of plant traits on belowground biomass would help understand the relationships between environment, plant traits and ecosystem properties.

Testing different traits:
Leaf resistance to fracture and leaf tensile strength:
These related traits indicate the carbon investment of the plant in the structural texture of the photosynthetic tissue (Cornelissen and Thompson 1997, Wright and Cannon 2001). Physically stronger leaves contribute to longer leaf lifespan and are better protected against abiotic (e.g. wind) and biotic (e.g. herbivory) damage. It could be expected that those traits effect decomposition rates and attenuate wave energy.

Leaf size:
Leaf size is the one-sided projection of the surface area of a leaf, expressed in mm^2 (Cornelissen et al. 2003). Smaller leaves were found to expand more rapidly than large-sized leaves and can thus contribute faster to positive net-photosynthesis (Moles and Westoby 2001). Ackerly and Reich (1999) found leaf size to be positively correlated with SLA and negatively with leaf lifespan and leaf nitrogen content. In the frame of this thesis, leaf size could contribute as surrogate for the hard trait leaf nitrogen-content, and is expected to affect decomposition rates negatively, as longer lived leaves are often rich in carbon, which decomposes less rapidly than nutrient rich, carbon poor leaves (Aerts and Chapin 2000).

References

Ackerly, D. D., C. A. Knight, S. B. Weiss, K. Barton, and K. P. Starmer. 2002. Leaf size, specific leaf area and microhabitat distribution of chaparral woody plants: contrasting patterns in species level and community level analyses. Oecologia **130**: 449-457.

Ackerly, D. D. and P. B. Reich. 1999. Convergence and correlations among leaf size and function in seed plants: A comparative test using independent contrasts. American Journal of Botany **86**: 1272-1281.

Ad-Hoc-AG Boden. 2005. Bodenkundliche Kartieranleitung. E. Schweitzerbart'sche Verlagsbuchhandlung, Hannover.

Adam, P. 1990. Saltmarsh ecology. Cambridge University Press, Cambridge.

Aerts, R. and F. S. I. Chapin. 2000. The mineral nutrition of wild plants revisited: A re-evaluation of processes and patterns. Advances of Ecological Research **30**: 1-67.

Ågren, G. 2004. The C : N : P stoichiometry of autotrophs - theory and observations. Ecology Letters **7**: 185-191.

Ågren, G. 2008. Stoichiometry and nutrition of plant growth in natural communities. Annual Review of Ecology, Evolution and Systematics **39**: 153-170.

Allen, S. E. e. 1989. Chemical analysis of ecological materials. Blackwell Scientific Publications, Oxford, England.

Andersson, T. 1997. Seasonal dynamics of biomass and nutrients in *Hepatica nobilis*. Flora **192**: 185-195.

Andratschke, S. 2009. Vegetationsökologische Untersuchungen und Management einer anthropogen beeinflussten Salzwiese sowie Leaf lifespan salztypischer Arten. Diploma thesis. University of Oldenburg, Oldenburg.

Arbuckle, J. L. 2007. Amos 16.0.1. Amos Development Corporation Spring House, PA, USA.

Armstrong, W., E. J. Wright, S. Lythe, and T. J. Gaynard. 1985. Plant zonation and the effects of the spring-neap tidal cycle on the soil aeration in a Humber salt marsh. Journal of Ecology **73**: 323-339.

Aronsson, A. and S. Elowson. 1980. Effects of irrigation and fertilization on mineral nutrients in Scots pine needles. Ecological Bulletins **32**: 219-228.

Austin, M. P. 1999. The potential contribution of vegetation ecology to biodiversity research. Ecography **22**: 465-484.

Austin, M. P. and T. M. Smith. 1989. A new model for the continuum concept. Vegetatio **83**: 35-47.

Backhaus, K., B. Erichson, W. Plinke, and R. Weiber. 2003. Multivariate Analysemethoden. Springer-Verlag, Berlin.

Bakker, J. P. 1989. Nature management by grazing and cutting. Kluwer Academic Publishers, Dordrecht.

Bakker, J. P., D. Bos, J. Stahl, Y. De Vries, and A. Jensen. 2003. Biodiversität und Landnutzung in Salzwiesen. Nova Acta Leopoldina NF 87 **328**: 163-194.

Bakker, J. P., J. Bunje, K. S. Dijkema, J. Frikke, N. Hecker, B. Kers, P. Körber, J. Kohlus, and M. Stock. 2005. Salt Marshes.*in* K. Essink, C. Dettman, H. Farke, K. Laursen, G. Lüerßen, H. Marencic, and W. Wiersinga, editors. Wadden Sea Quality Status Report 2004, Wadden Sea Ecosystem No. 19 - 2005. Common Wadden Sea Secretariat (CWSS), Wilhelmshaven.

Bakker, J. P., J. De Leeuw, K. S. Dijkema, P. C. Leendertse, H. H. T. Prins, and J. Rozema. 1993. Salt marshes along the coast of The Netherlands. Hydrobiologia **265**: 73.

Barkla, B. J. and O. Pantoja. 1996. Physiology of ion transport across the tonoplast of higher plants. Annual Review of Plant Physiology and Plant Molecular Biology **47**: 159-184.

Batten, G. D. and I. F. Wardlaw. 1987. Senescence and grain development in wheat plants grown with contrasting phosphorus regimes. Australian Journal of Plant Physiology **14**: 253-265.

Bazzaz, F. A. 1997. Allocation of resources in plants: state of the science and critical questions. Pages 1-37 *in* F. A. Bazzaz and J. Grace, editors. Plant Resource Allocation Academic Press, San Diego, CA.

Bazzaz, F. A. and E. G. Reekie. 1985. The meaning and measurement of reproductive effort in plants. Pages 373-387 *in* J. White, editor. Studies on Plant Demographie. Academic Press, London.

Beeftink, A. 1966. Vegetation and habitat of salt marshes and beach plains in S.W. Netherlands. Wentia **15**: 83-108.

Berendse, F. and F. Möller. 2009. Effects of competition on root-shoot allocation in *Plantago lanceolata* L.: adaptive plasticity or ontogenetic drift? Plant Ecology **201**: 567-573.

Bertness, M. D. 1991. Zonation of *Spartina patens* and *Spartina alterniflora* in a New England salt marsh. Ecology **72**: 138-148.

Bertness, M. D. and A. M. Ellison. 1987. Determinants of pattern in a New England salt marsh plant community. Ecological Monographs **57**: 129-147.

Bertness, M. D. and G. H. Leonard. 1997. The role of positive interactions in communities: lessons from intertidal habitats. Ecology **78**: 1976-1989.

Biere, A. 1995. Genotypic and plastic variation in plant size: effects on fecundity and allocation patterns in *Lychnis flos-cuculi* along a gradient of natural soil fertility. Journal of Ecology **83**: 629-642.

Blew, J., K. Günther, K. Laursen, M. van Roomen, P. Südbeck, K. Eskildsen, P. Potel, and H.-U. Rösner. 2005. Overview on trend and numbers of migratory waterbirds in the Wadden Sea 1980-2000 *in* J. Blew and P. Südbeck, editors. Migratory water birds in the Wadden Sea 1980-2000. Common Wadden Sea Secretariat (CWSS) Wilhelmshaven, Germany.

Bockelmann, A. C., J. P. Bakker, R. Neuhaus, and J. Lage. 2002. The relation between vegetation zonation, elevation and inundation frequency in a Wadden Sea salt marsh. Aquatic Botany **73**: 211-221.

Bockelmann, A. C. and R. Neuhaus. 1999. Competitive exclusion of *Elymus athericus* from a high stress habitat in a European salt marsh. Journal of Ecology **87**: 503-513.

Borowitzka, L. 1981. Solute accumulation and regulation of cell water activity. Pages 97-104 *in* L. G. Palek and D. Aspinall, editors. Physiology and biochemistry of drought resistance in plants. Academic Press, Sydney.

Bouma, T. J., M. B. De Vries, E. Low, G. Peralta, I. C. Tánczos, J. van de Koppel, and P. M. J. Herman. 2005. Trade-offs related to ecosystem engineering: a case study on stiffness of emerging macrophytes. Ecology **86**: 2187-2199.

Bouma, T. J., B. P. Koustaal, M. van Dongen, and K. L. Nielsen. 2001. Coping with low nutrient availability and inundation: root growth responses of three halophytic grass species from different elevations along a flooding gradient. Oecologia **126**: 472-481.

Brandt, A. C. and A. Wollesen. 2009. Tourism and Recreation.*in* H. Marencic and J. de Vlas, editors. Wadden Sea Ecosystem No. 25, Quality Status Report 2009, Thematic Report No. 3.4. Common Wadden Sea Secretariat (CWSS), Trilateral Monitoring and Assessment Group (TMAG). Wilhelmshaven, Germany.

Brenchley, W. E. 1916. The effect of the concentration of the nutrient solution on the growth of barley and wheat in water cultures. Annals of Botany **30**: 77-91.

Brouwer, R. 1962a. Distribution of dry matter in plants. Netherlands Journal of Agricultural Sciences **10**: 399-408.

Brouwer, R. 1962b. Nutritive influences on the distribution of dry matter in the plant. Netherlands Journal of Agricultural Sciences **10**: 361-376.

Brown, J. H., J. F. Gillooly, A. P. Allen, V. M. Savage, and G. B. West. 2004. Toward a metabolic theory in ecology. Ecology **85**: 1771-1789.

Brown, S. L. 1998. Sedimentation on a Humber saltmarsh. Pages 69-83 *in* K. S. Black, D. M. Paterson, and A. Cramp, editors. Sedimentary Processes in the Intertidal Zone. Geological Society, Special Publications, London.

Browne, M. and R. Cudeck. 1993. Alternative ways of assessing equation model fit. Pages 136-162 *in* K. A. Boollen and J. S. Long, editors. Testing Structural Equation Models, Newbury Parks.

Burke, M. J. W. and J. P. Grime. 1996. An experimental study of plant community invasibility. Ecology **77**: 776-790.

Calder, I. W. A. 1996. Size, function, and life history. Dover, New York.

Calinski, T. and J. Harabasz. 1974. A Dendrite Method for Cluster Analysis. Communications in Statistics **3**: 1-27.

Campbell, B. D. and J. P. Grime. 1992. An experimental test of plant strategy theory. Ecology **73**: 15-29.

Cargill, S. M. and R. L. Jefferies. 1984. Nutrient limitation of primary production in a sub-arctic salt marsh Journal of Applied Ecology **21**: 657-668.

Chapin, F. S. I. 1980. The mineral nutrition of wild plants. Annual Review of Ecology and Systematics **11**: 233-260.

Chapin, F. S. I., K. Autumn, and F. Pugnaire. 1993. Evolution of suites of traits in response to environmental stress. American Naturalist **142** (Suppl): S78-S92.

Chapin, F. S. I., M. S. Bret-Harte, S. E. Hobbie, and Z. Hailin. 1996. Plant functional types as predictors of transient responses of arctic vegetation to global change. Journal of Vegetation Science **7**: 347-358.

Chapin, F. S. I., P. A. Matson, and H. A. Mooney. 2002. Principles of Terrestrial Ecosystem Ecology. Springer, New York.

Chapin, F. S. I. and G. R. Shaver. 1989. Differences in growth and nutrient use among arctic plant growth forms. Functional Ecology **3**: 73-80.

Chapin, F. S. I., B. H. Walker, R. J. Hobbs, D. U. Hooper, J. H. Lawton, O. E. Sala, and D. Tilman. 1997. Biotic control over the functioning of ecosystems. Science **277**: 500-504.

Chapin, F. S. I., E. S. Zavaleta, V. T. Eviner, R. L. Naylor, P. M. Vitousek, H. L. Reynolds, D. U. Hooper, S. Lavorel, O. E. Sala, S. E. Hobbie, M. C. Mack, and S. Díaz. 2000. Consequences of changing biodiversity. Nature **405**: 234-242.

Chapman, V. J. 1974. Salt marshes and salt deserts of the world. J. Cramer, Lehre.

Cheeseman, J. M. 1993. Plant growth modelling without integrating mechanisms. Plant, Cell and Environment **16**: 137-147.

Chidumayo, E. N. 1994. Phenology and nutrition of miombo woodland trees in Zambia. Trees **9**: 67-72.

Choler, P. 2005. Consistent shifts in Alpine plant traits along a mesotopographical gradient. Arctic, Antarctic and Alpine Research **37**: 444-453.

Christensen, N. L., A. M. Bartuska, J. H. Brown, S. Carpenter, C. Dantonio, R. Francis, J. F. Franklin, J. A. MacMahon, R. F. Noss, D. J. Parsons, C. H. Peterson, M. G. Turner, and R. G. Woodmansee. 1996. The report of the ecological society of America committee on the scientific basis for ecosystem management. Ecological Applications **6**: 665-691.

Cooper, A. 1982. The effects of salinity and waterlogging on the growth and cation uptake of salt marsh plants. New Phytologist **90**: 263-275.

Cornelissen, J. H. C., P. C. Diez, and R. Hunt. 1996. Seedling growth, allocation and leaf attributes in a wide range of woody plant species and types. Journal of Ecology **84**: 755-765.

Cornelissen, J. H. C., S. Lavorel, E. Garnier, S. Diaz, N. Buchmann, D. E. Gurvich, P. B. Reich, H. ter Steege, H. D. Morgan, M. G. A. van der Heijden, J. G. Pausas, and H. Poorter. 2003. A handbook of protocols for standardised and easy measurement of plant functional traits worldwide. Australian Journal of Botany **51**: 335-380.

Cornelissen, J. H. C. and K. Thompson. 1997. Functional leaf attributes predict litter decomposition rates in herbaceous plants. New Phytologist **135**: 109-114.

Craine, J. M., W. G. Lee, W. J. Bond, R. J. Williams, and L. C. Johnson. 2005. Environmental constraints on a global relationship among leaf and root traits of grasses. Ecology **86**: 12-19.

Cunningham, S., B. Summerhayes, and M. Westoby. 1999. Evolutionary divergences in leaf structure and chemistry, comparing rainfall and soil nutrient gradients. Ecological Monographs **69**: 569-588.

Darwin, C. 1859. On the origin of species. John Murray.

Davy, A. J., C. S. B. Costa, A. R. Yallop, A. M. Proudfoot, and M. F. Mohamed. 2000. Biotic interactions in plant communities in saltmarshes. Linnean Society of London by Forrest Text, Tresaith, Ceridigion.

De Deyn, G. B., J. H. C. Cornelissen, and R. D. Bardgett. 2008. Plant functional traits and soil carbon sequestration in contrasting biomes. Ecology Letters **11**: 516-531.

de Graaf, M. C. C., R. Bobbink, J. G. M. Roelofs, and P. J. M. Verbeek. 1998. Differential effects of ammonium and nitrate on three heathland species. Plant Ecology **135**: 185-196.

De Leeuw, J., H. Olff, and J. P. Bakker. 1990. Year-to-year variation in peak above-ground biomass of six salt-marsh angiosperm communities as related to rainfall deficit and inundation frequency. Aquatic Botany **36**: 139-151.

Deutscher Wetterdienst. 2009. Mean climate values for the period 1961 to 1990.[http://www.dwd.de/bvbw/appmanager/bvbw/dwdwwwDesktop?_nfpb=true&_pageLabel =_dwdwww_menu2_leistungen_a-z_freiemetinfos&T115202758871200642573928gsbDocumentPath=Navigation%2FOeffentlich keit%2FKlima__Umwelt%2FKlimadatenzentren%2FNKDZ%2Fkldaten__akt%2Fausgabe__mi ttelwerte__node.html%3F__nnn%3Dtrue] Accessed: 06-04-2009.

Díaz, S. and M. Cabido. 1997. Plant functional types and ecosystem function in relation to global change. Journal of Vegetation Science **8**: 463-474.

Díaz, S. and M. Cabido. 2001. Vive la difference: plant functional diversity matters to ecosystem processes. Trends in Ecology & Evolution **16**: 646-655.

Díaz, S., J. Fargione, I. F. S. Chapin, and D. Tilman. 2006. Biodiversity loss threatens human well-being. PLoS Biology **4**: 1300-1305.

Díaz, S., S. Lavorel, F. De Bello, F. Quétier, K. Grigulis, and M. T. Robson. 2007. Incorporating plant functional diversity effects in ecosystem service assessments. Proceedings of the National Academy of Sciences **104**: 20684-20689.

Dolédec, S., D. Chessel, C. J. F. Ter Braak, and S. Champely. 1996. Matching species traits to environmental variables: a new three-table ordination method. Environmental and Ecological Statistics **3**: 143-166.

Dormann, C. F., R. Van der Wal, and J. P. Bakker. 2000. Competition and herbivory during salt marsh succession: the importance of forb growth strategy. Journal of Ecology **88**: 571-583.

Dray, S., D. Chessel, and J. Thioulouse. 2003. Co-inertia anaylsis and the linking of ecological data tables. Ecology **84**: 3078-3089.

Dray, S. and P. Legendre. 2008. Testing the species traits-environment relationships: the fourth-corner problem revisited. Ecology **89**: 3400-3412.

Dray, S., N. Pettorelli, and D. Chessel. 2002. Matching data sets from two different spatial samples. Journal of Vegetation Science **13**: 867-874.

Duckworth, J. C., M. Kent, and P. M. Ramsay. 2000. Plant functional types: an alternative to taxonomic plant community description in biogeography? Progress in Physical Geography **24**: 515-542.

Eckstein, R. L. and P. S. Karlsson. 1997. Above-ground growth and nutrient use by plants in a subarctic environment: effects of habitat, life-form and species. Oikos **79**: 311-324.

Egan, T. P. and I. A. Ungar. 2001. Competition between *Salicornia europeae* and *Atriplex prostrata* (Chenopodiaceae) along an experimental salinity gradient. Wetlands Ecology and Management **9**: 457-461.

Egnér, H., H. Riehm, and W. R. Domingo. 1960. Untersuchungen über die Bodenanalyse als Grundlage für die Beurteilung des Nährstoffzustandes des Bodens II. Chemische Extraktionsmethoden zur Phosphor- und Kaliumbestimmung. Kungl. Lantbrukshögskolans Annualer **26**: 199-215.

Elser, J. J., M. E. S. Bracken, E. E. Cleland, D. S. Gruner, W. S. Harpole, H. Hillebrand, J. T. Ngai, E. W. Seabloom, J. B. Shurin, and J. E. Smith. 2007. Global analysis of nitrogen and phosphorus limitation of primary producers in freshwater, marine and terrestrial ecosystems. Ecology Letters **10**: 1135-1142.

Elser, J. J., W. F. Fagan, R. F. Denno, D. R. Dobberfuhl, A. Folarin, A. Huberty, S. Interlandi, S. S. Kilham, E. McCauley, K. L. Schulz, E. H. Siemann, and R. W. Sterner. 2000a. Nutritional constraints in terrestrial and freshwater food webs. Nature **408**: 578-580.

Elser, J. J. and A. Hamilton. 2007. Stoichiometry and the new biology - The future is now. PLoS Biology **5**: 1403-1405.

Elser, J. J., R. W. Sterner, E. Gorokhova, W. F. Fagan, T. A. Markow, J. B. Cotner, J. F. Harrison, S. E. Hobbie, G. M. Odell, and L. J. Weider. 2000b. Biological stoichiometry from genes to ecosystems. Ecology Letters **3**: 540-550.

Engels, J. G. and K. Jensen. 2010. Role of biotic interactions and physical factors in determining the distribution of marsh species along an estuarine salinity gradient. Oikos **119**: 679-685.

Enquist, B. J. and K. J. Niklas. 2001. Invariant scaling relations across tree-dominated communities Nature **410**: 655-660.

Enquist, B. J. and K. J. Niklas. 2002. Global allocation rules for patterns of biomass partitioning in seed plants. Science **295**:1517-1520.

Epstein, E. 1972. Mineral nutrition of plants: principles and perspectives. Wiley, New York.

Ernst, W. H. O. 1983. Ökologische Anpassungsstrategien an Bodenfaktoren. Berichte der Deutschen Botanischen Gesellschaft **96**: 49-71.

Esselink, P., J. Petersen, S. Arens, J. P. Bakker, J. Bunje, K. S. Dijkema, N. Hecker, U. Hellwik, A.-V. Jensen, A. S. Kers, P. Körber, E. J. Lammerts, M. Stock, R. M. Veeneklaas, M. Vreeken, and M. Wolters. 2009. Salt Marshes - Thematic Report No. 8.*in* H. Marencic, editor. Wadden Sea Quality Status Report 2009, Wadden Sea Ecosystem No. 25 - 2009. Common Wadden Sea Secretariat (CWSS), Trilateral Monitoring and Assessment Group (TMAG), Wilhelmshaven.

Etherington, J. R. 1975. Environment and Plant Ecology. John Wiley & Sons, London.

Eviner, V. T. 2004. Plant traits that influence ecosystem processes vary independently among species. Ecology **85**: 2215-2229.

Fenner, M. 1986. The allocation of minerals to seeds in *Senecio vulgaris* plants subjected to nutrient shortage. Journal of Ecology **74**:385-392.

Flexas, J., J. Bota, F. Loreto, G. Cornic, and T. D. Sharkey. 2004. Diffusive and metabolic limitations to photosynthesis under drought and salinity in C_3 plants. Plant Biology **6**: 269-279.

Flowers, F. J. 1975. Halophytes. Pages 309-334 *in* D. A. Baker and J. L. Hall, editors. Ion Transport in Plant Cells and Tissues. North Holland.

Flowers, T. J. and T. D. Colmer. 2008. Salinity tolerance in halophytes. New Phytologist **179**: 945-963.

Föhse, D., N. Claassen, and A. Jungk. 1988. Phosphorus efficiency of plants Plant and Soil **110**: 101-109.

Fonseca, C. R., J. M. Overton, B. Collins, and M. Westoby. 2000. Shifts in traits-combinations along rainfall and phosphorus gradients. Journal of Ecology **88**: 964-977.

Foth, H. D. and B. G. Ellis. 1996. Soil Fertility. 2nd edition. CRC Press LLB, Boca Raton, Florida, USA.

Freschet, G. T., J. H. C. Cornelissen, R. S. P. van Logtestijn, and R. Aerts. 2010. Evidence of the 'plant economics spectrum' in a subarctic flora. Journal of Ecology **98**: 362-373.

Frost, P. C., M. A. Evans-White, Z. V. Finkel, T. C. Jensen, and V. Matzek. 2005. Are you what you eat? Physiological constraints on organismal stoichiometry in an elementally imbalanced world. Oikos **109**: 18-28.

Garnier, E., P. Cordonnier, J.-L. Guillerm, and L. Sonié. 1997. Specific leaf area and nitrogen concentration in annual and perennial species growing in Mediterranean old-fields. Oecologia **111**: 490-498.

Garnier, E., J. Cortez, G. Billès, M.-L. Navas, C. Roumet, M. Debussche, G. Laurent, A. Blanchard, D. Aubry, A. Bellmann, C. Neill, and J.-P. Toussaint. 2004. Plant functional markers capture ecosystem properties during secondary succession. Ecology **85**: 2630-2637.

Garnier, E., S. Lavorel, P. Ansquer, H. Castro, P. Cruz, J. Dolezal, O. Eriksson, C. Fortunel, H. Freitas, C. Golodets, K. Grugulis, C. Jouany, E. Kazakou, J. Kigel, M. Kleyer, V. Lehsten, J. Lepš, T. Meier, R. Pakeman, M. Papadimitriou, V. P. Papanastasis, H. Quested, F. Quétier, M. Robson, C. Roumet, G. Rusch, C. Skarpe, M. Sternberg, J.-P. Theau, A. Thébault, D. Vile, and M. Zarovali. 2007. Assessing the effects of land-use change on plant traits, communities and ecosystem functioning in grasslands: a standardized methodology and lessons from an application to 11 European sites. Annals of Botany **99**: 967-985.

Garnier, E., B. Shipley, C. Roumet, and G. Laurent. 2001. A standardized protocol for the determination of specific leaf area and leaf dry matter content. Functional Ecology **15**: 688-695.

Gaudet, C. and P. A. Keddy. 1988. A comparative approach to prediciting competitive ability from plant traits. Nature **334**: 242-243.

Geber, M. A. and L. R. Griffen. 2003. Inheritance and natural selection on functional traits. International Journal of Plant Science **164**: S21-S43.

Geider, R. J. and J. La Roche. 2002. Redfield revisited: variability of C:N:P in marine microalgae and its biochemical basis. European Journal of Phycology **37**: 1-17.

Gitay, H. and I. R. Noble. 1997. What are functional types and how should we seek them? Pages 3-19 *in* T. M. Smith, H. H. Shugart, and F. I. Woodward, editors. Plant functional types: their relevance to ecosystem properties. Cambridge University Press, Cambridge.

Golodets, C., M. Sternberg, and J. Kigel. 2009. A community-level test of the leaf-height-seed ecology strategy scheme in relation to grazing conditions. Journal of Vegetation Science **20**: 392-402.

González, A. L., J. S. Kominoski, M. Danger, S. Ishida, N. Iwai, and A. Rubach. 2010. Can ecological stoichiometry help explain patterns of biological invasions? Oikos **119**: 779-790.

Gordon, W. S. and R. B. Jackson. 2000. Nutrient concentrations in fine roots. Ecology **81**: 275-280.

Gorham, J., L. Hughes, and R. G. Wyn Jones. 1981. Low-molecular-weight carbohydrates in some salt-stressed plants. Physiologia Plantarum **53**: 27-33.

Gorham, J., L. L. Hughes, and R. G. Wyn Jones. 1980. Chemical composition of salt-marsh plants from Ynys Môn (Anglesey): the concept of physiotypes. Plant, Cell and Environment **3**: 309-318.

Grace, J. 2006. Structural Equation Modeling and Natural Systems. Cambridge University Press, Cambridge.

Grace, J. B. and B. H. Pugesek. 1997. A structural equation model of plant species richness and its application to a coastal wetland. The American Naturalist **149**: 436-460.

Grace, J. B. and B. H. Pugesek. 1998. On the use of path analysis and related procedures for the investigation of ecological problems. The American Naturalist **152**: 151-159.

Gray, A. J. and R. G. H. Bunce. 1972. The ecology of Morecambe Bay VI. Soils and vegetation of the salt marshes: a multivariate approach. Journal of Applied Ecology **9**: 221-234.

Grigal, D. F., L. F. Ohmann, and R. B. Brander. 1976. Seasonal dynamics of tall shrubs in Northeastern Minnesota: biomass and nutrient element changes Forest Science **22**: 195-208.

Grime, J. P. 1973. Competition and diversity in herbaceous vegetation - a reply. Nature **244**: 310-311.

Grime, J. P. 1974. Vegetation classification by reference to strategies. Nature **250**: 26-31.

Grime, J. P. 1979. Evidence for the existence of three primary strategies in plants and its relevance to ecological and evolutionary theory. American Naturalist **111**: 1169-1194.

Grime, J. P. 1998. Benefits of plant diversity to ecosystems: immediate, filter and founder effects. Journal of Ecology **86**: 902-910.

Grime, J. P. 2002. Plant strategies, vegetation processes, and ecosystem properties. John Wiley & Sons, Ltd., Chichester.

Groenendijk, A. M. 1985. Ecological consequences of tidal management for the salt-marsh vegetation. Vegetatio **62**: 415-424.

Groenendijk, A. M. and M. A. Vink-Lievaart. 1987. Primary production and biomass on a Dutch salt marsh: emphasis on the below-ground component. Vegetatio **70**: 21-27.

Grubb, P. J. 1985. Plant populations and vegetation in relation to habitat, disturbance and competition: problems of generalization. Pages 595-621 *in* J. White, editor. The Population Structure of Vegetation. Junk, Dordrecht.

Güsewell, S. 2004. N:P ratios in terrestrial plants: variation and functional significance. New Phytologist **164**: 243-266.

Güsewell, S. and W. Koerselman. 2002. Variation in nitrogen and phosphorus concentrations of wetland plants. Perspectives in Plant Ecology, Evolution and Systematics **5**: 37-61.

Gutknecht, J. and J. Dainty. 1969. Ionic relationships of marine algae. Oceanography and Marine Biology-An Annual Review **6**: 163-200.

Guy, R. D. and D. M. Reid. 1986. Photosynthesis and the influence of CO_2-enrichment on $\delta^{13}C$ values in a C_3 halophyte. Plant, Cell and Environment **9**: 65-72.

Hammer, Ø., D. A. T. Harper, and P. D. Ryan. 2001. PAST: Paleontological Statistics Software Package for Education and Data Analysis. Palaeontologia Electronica **4**: 9.

Handreck, K. A. 1997. Phosphorus requirements of Australian native plants. Australian Journal of Soil Research **35**: 241-290.

Hasegawa, P. M., R. A. Bressan, J. K. Zhu, and H. J. Bohnert. 2000. Plant cellular and molecular responses to high salinity. Annual Review of Plant Physiology and Plant Molecular Biology **51**: 463-499.

Havill, D. C., A. Ingold, and J. Pearson. 1985. Sulphide tolerance in coastal halophytes. Vegetatio **62**: 279-285.

Hector, A. and R. Bagchi. 2007. Biodiversity and ecosystem multifunctionality. Nature **448**: 188-191.

Hemminga, M. A. and G. J. C. Buth. 1991. Decomposition in salt marsh ecosystems of the S.W. Netherlands: The effects of biotic and abiotic factors. Vegetatio **92**: 73-83.

Hemmingsen, A. M. 1960. Energy metabolism as related to body size and respiratory surfaces, and its evolution. Reports of the Steno Memorial Hospital and Nordisk Insulin Laboratorium **9**: 6-110.

Henley, W. J., S. T. Lindley, G. Levavasseur, C. B. Osmond, and J. Ramus. 1992. Photosynthetic response of *Ulva rotundata* to light and temperature during emersion on an intertidal sand flat. Oecologia **89**: 516-523.

Hessen, D. O., G. I. Agren, T. R. Anderson, J. J. Elser, and P. C. De Ruiter. 2004. Carbon sequestration in ecosystems: The role of stoichiometry. Ecology **85**: 1179-1192.

Heydemann, B. 1981. Ecology of arthropods of the lower salt marsh Pages 35 – 57 *in* N. Dankers, H. Kühl, and W. J. Wolff, editors. Invertebrate fauna of the Wadden Sea. Report 4 of the Wadden Sea Working Group. Balkema, Rotterdam.

Hillebrand, H., E. T. Borer, M. E. S. Bracken, B. J. Cardinale, J. Cebrian, J. J. Elser, D. S. Gruner, W. S. Harpole, J. T. Ngai, S. Sandin, E. W. Seabloom, J. B. Shurin, J. E. Smith, and M. D. Smith. 2009. Herbivore metabolism and stoichiometry each constrain herbivory at different organizational scales across ecosystems. Ecology Letters **12**: 516-527.

Hodgson, J. G., P. J. Wilson, R. Hunt, J. P. Grime, and K. Thompson. 1999. Allocating C-S-R plant functional types: a soft approach to a hard problem. Oikos **85**: 282-294.

Hooper, D. U., J. J. Ewel, A. Hector, P. Inchausti, S. Lavorel, D. Lodge, M. Loreau, S. Naeem, B. Schmid, H. Setälä, A. J. Symstad, J. Vandermeer, and W. D. A. 2005. Effects of biodiversity on ecosystem functioning: a consensus of current knowledge and needs for further research. Ecological Monographs **75**: 3-35.

Howarth, R. W. 1988. Nutrient limitation of net primary production in marine ecosystems. Annual Review of Ecology and Systematics **19**: 89-110.

Howes, B. L., J. W. H. Dacey, and D. D. Goehringer. 1986. Factors controlling the growth form of *Spartina alterniflora*: feedbacks between above-ground production, sediment oxidation, nitrogen and salinity. Journal of Ecology **74**: 881-898.

Janiesch, P. 1991. Oberirdische Biomasseproduktion und Mineralstoffhaushalt von Salzwiesen der niedersächsischen Küste. Drosera **1/2**: 127-138.

Jardim, A. V. F. and M. A. Batalha. 2008. Can we predict dispersal guilds based on the leaf-height-seed scheme in a disjunct cerrado woodland? Brazilian Journal of Biology **68**: 553-559.

Jeffrey, D. W. 1964. The formation of polyphosphate in Banksia ornata, an Australian heath plant. Australian Journal of Biological Science **17**: 845-854.

Jensen, A. 1985. On the ecophysiology of *Halimione portulacoides*. Vegetatio **62**: 231-240.

Kazakou, E., D. Vile, B. Shipley, C. Gallet, and E. Garnier. 2006. Co-variations in litter decomposition, leaf traits and plant growth in species from a Mediterranean old-field succession. Functional Ecology **20**: 21-30.

Keddy, P. A. 1992. A pragmatic approach to functional ecology. Functional Ecology **6**: 621-626.

Kerkhoff, A. J., W. F. Fagan, J. J. Elser, and B. J. Enquist. 2006. Phylogenetic and growth form variation in the scaling of nitrogen and phosphorus in the seed plants. The American Naturalist **168**: 103-122.

Kiehl, K. 1997. Vegetationsmuster in Vorlandsalzwiesen in Abhängigkeit von Beweidung und abiotischen Standortbedingungen. Dissertation. University of Kiel, Kiel.

Kiehl, K., P. Esselink, and J. P. Bakker. 1997. Nutrient limitation and plant species composition in temperate salt marshes. Oecologia **111**: 325-330.

Kingsolver, J. G. and D. W. Schemske. 1991. Path analyses of selection. Trends in Ecology & Evolution **6**: 276-280.

Kinzel, H. 1982. Pflanzenökologie und Mineralstoffwechsel. Ulmer Verlag, Stuttgart.

Kirk, J. T. O. and R. A. E. Tilney-Bassett. 1978. The Plastids. 2nd edition. Elsevier, Amsterdam.

Klinkhamer, P. G. L., T. J. de Jong, and E. Meelis. 1990. How to test for proportionality in the reproductive effort of plants. The American Naturalist **135**: 291-300.

Knecht, M. F. and A. Göransson. 2004. Terrestrial plants require nutrients in similar proportions. Tree Physiology **24**: 447-460.

Knevel, I. C., R. M. Bekker, D. Kunzmann, M. Stadler, and K. Thompson, editors. 2005. The LEDA Traitbase - Collecting and Measuring Standards of Life-history Traits of the Northwest European Flora. LEDA Traitbase project, University of Groningen, Community and Conservation Ecology group Groningen, NL.

Knox, E. B. and J. D. Palmer. 1995. Chloroplast DNA variation and the recent radiation of the giant senecios (Asteraceae) on the tall mountains of eastern Africa. Proceedings of the National Academy of Sciences of the United States of America **92**: 10349-10353.

Koffijberg, K., J. Blew, K. Eskildsen, K. Günther, B. Koks, K. Laursen, L.-M. Rasmussen, P. Südbeck, and P. Potel. 2003. High tide roosts in the Wadden Sea: A review of bird distribution, protection regimes and potential sources of anthropogenic disturbance. A report of the Wadden Sea Plan Project 34. Wadden Sea Ecosystems No. 16. Common Wadden Sea Secretariat, Trilateral Monitoring and Assessment Group, JMMB, Wilhelmshaven, Germany.

Körner, C. 1994. Biomass fractionation in plants: a reconsideration of definitions based on plant functions. Pages 173-185 *in* J. Roy and E. Garnier, editors. A whole plant perspective on carbon–nitrogen interactions. SPB Academic Publishing, The Hague.

Koyro, H.-W. 2006. Effect of salinity on growth, photosynthesis, water relations and solute composition of the potential cash crop halophyte *Plantago coronopus* (L.) Environmental and Experimental Botany **56**: 136-146.

Kozlowski, T. T. 1984. Plant responses to flooding of soil. Bioscience **34**: 162-167.

Kramer, P. J. and T. T. Kozlowski. 1979. Physiology of woody plants. Academic, New York.

Krebs, C. J. 1989. Ecological Methodology. Harper & Row Publishers, New York.

Kuhbier, H. 1975. Das Pflanzenkleid der Insel Mellum.*in* P. Blasyk, editor. Naturschutzgebiete im Oldenburger Land. Heinz Holzberg Verlag, Oldenburg.

Kühner, A. and M. Kleyer. 2008. A parsimonious combination of functional traits predicting plant response to disturbance and soil fertility. Journal of Vegetation Science **19**: 681-692.

Kuijper, D. P. J., J. Dubbeld, and J. P. Bakker. 2005. Competition between two grass species with and without grazing over a productivity gradient. Plant Ecology **179**: 237-246.

Künnemann, T.-D. and G. Gad. 1997. Überleben zwischen Land und Meer. Isensee, Oldenburg.

Laughlin, D. C., J. J. Leppert, M. M. Moore, and C. H. Sieg. 2010. A multi-trait test of the leaf-height-seed plant strategy scheme with 133 species from a pine forest flora. Functional Ecology **24**: 493-501.

Lavorel, S., S. Díaz, J. H. C. Cornelissen, E. Garnier, S. P. Harrison, S. McIntyre, J. G. Pausas, N. Pérez-Harguindeguy, C. Roumet, and C. Urcelay. 2007. Plant functional types: Are we getting any closer to the holy grail? Series: Global change - The IGBP series edition. Springer, Berlin.

Lavorel, S. and E. Garnier. 2002. Predicting changes in community composition and ecosystem functioning from plant traits: revisiting the Holy Grail. Functional Ecology **16**: 545-556.

Lavorel, S., S. McIntyre, J. Landsberg, and T. D. A. Forbes. 1997. Plant functional classifications: from general groups to specific groups based on response of disturbance. Trends in Ecology and Evolution **12**: 474-478.

Lavorel, S. and D. M. Richardson. 1999. Diversity, stability and conservation of mediterranean-type ecosystems in a changing world: an introduction. Diversity and Distributions **5**: 1-2.

Leeuw, J. d., H. Olff, and J. P. Bakker. 1990. Year-to-year variation in peak above-ground biomass of six salt-marsh angiosperm communities as related to rainfall deficit and inundation frequency. Aquatic Botany **36**: 139-151.

Legendre, P., R. Galzin, and M. L. Harmelin-Vivien. 1997. Relating behavior to habitat: solutions to the fourth-corner problem. Ecology **78**: 574-562.

Legendre, P. and L. Legendre. 1998. Numerical Ecology. 2nd edition. Elsevier.

Lenssen, G. M., J. Lamers, M. Stroetenga, and J. Rozema. 1993. Interactive effects of atmospheric CO_2 enrichment, salinity and flooding on growth of C_3 (*Elymus athericus*) and C_4 (*Spartina anglica*) salt marsh species. Vegetatio **104/105**: 379-388.

Lenssen, G. M., W. E. van Duin, P. Jak, and J. Rozema. 1995. The response of *Aster tripolium* and *Puccinellia maritima* to atmospheric carbon dioxide enrichment and their interactions with flooding and salinity. Aquatic Botany **50**: 181-192.

Leuschner, C., S. Landwehr, and U. Mehlig. 1998. Limitation of carbon assimilation of intertidal *Zostera noltii* and *Z. marina* by desiccation at low tide. Aquatic Botany **62**: 171-176.

Lewis, J. D. and R. T. Koide. 1990. Phosphorus supply, mycorrhizal infection and plant offspring vigour. Functional Ecology **4**: 695-702.

Leyer, I. and K. Wesche. 2007. Multivariate Statistik in der Ökologie. Springer Verlag, Berlin.

Liebig, J. 1840. Die organische Chemie in ihrer Anwendung auf Agricultur und Physiologie. Friedrich Vieweg und Sohn, Braunschweig, Germany.

Liebig, J. 1855. Die Grundsätze der Agricultur-Chemie mit Rücksicht auf die in England angestellten Untersuchungen. 2nd edition. Friedrich Vieweg und Sohn, Braunschweig, Germany.

Loreau, M. 1998. Biodiversity and ecosystem functioning: A mechanistic model. Proceedings of the National Academy of Sciences **95**: 5632-5636.

Loreau, M., S. Naeem, P. Inchausti, J. Bengtsson, J. P. Grime, A. Hector, D. U. Hooper, M. A. Huston, D. Raffaelli, B. Schmid, D. Tilman, and D. A. Wardle. 2001. Biodiversity and ecosystem functioning: current knowledge and future challenges. Science **294**: 804-808.

Lucassen, E. C. H. E. T., R. Bobbink, A. J. P. Smolders, P. J. M. van der Ven, L. P. M. Lamers, and J. G. M. Roelofs. 2002. Interactive effects of low pH and high ammonium levels responsible for the decline of *Cirsium dissectum* (L.) Hill. Plant Ecology **165**: 45-52.

Mardia, K. V. 1970. Measures of multivariate skewness and kurtosis with applications. Biometrika **57**: 519-530.

Mardia, K. V. 1974. Applications of some measures of multivariate skewness and kurtosis in testing normality and robustness studies. Sankhyā - The Indian Journal of Statistics, Series B. **36**: 115-128.

Matzek, V. and P. M. Vitousek. 2009. N:P stoichiometry and protein:RNA ratios in vascular plants: an evaluation of the growth-rate hypothesis. Ecology Letters **12**: 765-771.

Maximov, N. A. 1929. The plant in relation to water. Allen and Unwin, London.

McCarthy, M. C. and B. J. Enquist. 2007. Consistency between an allometric approach and optimal partitioning theory in global patterns of plant biomass allocation. Functional Ecology **21**: 713-720.

McConnaughay, K. D. M. and J. S. Coleman. 1999. Biomass allocation in plants: Ontogeny or optimality? A test along three resource gradients. Ecology **80**: 2581-2593.

McCune, B. and J. B. Grace. 2002. Structural Equation Modeling. Pages 233-256 *in* B. McCune and J. B. Grace, editors. Analysis of Ecological Communities. MjM Software Design, Gleneden Beach, Oregon.

McGill, B. J., B. J. Enquist, E. Weiher, and M. Westoby. 2006. Rebuilding community ecology from functional traits. Trends in Ecology & Evolution **21**: 178-184.

McGroddy, M. E., T. Daufresne, and L. O. Hedin. 2004. Scaling of C : N : P stoichiometry in forests worldwide: Implications of terrestrial Redfield-type ratios. Ecology **85**: 2390-2401.

McNaughton, S. J. 1977. Diversity and stability of ecological communities: A comment on the role of empiricism in ecology. The American Naturalist **111**: 515-525.

Minden, V., S. Andratschke, J. Spalke, H. Timmermann, and M. Kleyer. unpubl. Trait-environmental concepts are not explicitly convertible to salt marshes

Minden, V., K. J. Hennenberg, S. Porembski, and H. J. Boehmer. 2010a. Invasion and management of alien *Hedychium gardnerianum* (kahili ginger, Zingiberaceae) alter plant species composition of a montane rainforest on the island of Hawai'i. Plant Ecology **206**: 321-333.

Minden, V., J. Jacobi, S. Porembski, and H. J. Boehmer. 2010b. Effects of invasive alien kahili ginger (*Hedychium gardnerianum*) on native plant species regeneration in a Hawaiian rainforest. Applied Vegetation Science **13**: 5-14.

Minden, V. and M. Kleyer. unpubl. Allometric relationships between plant biomass traits and carbon/nitrogen ratios are keystone response and effect traits determining aboveground biomass of salt marshes.

Mitchell, R. J. 1992. Testing evolutionary and ecological hypotheses using path analysis and structural equation modeling. Functional Ecology **6**: 123-129.

Moles, A. T. and M. Westoby. 2001. Do small leaves expand faster than large leaves, and do shorter expansion times reduce herbivore damage? Oikos **90**: 517-524.

Moles, A. T. and M. Westoby. 2006. Seed size and plant strategy across the whole life cycle. Oikos **133**: 91-105.

Morton, A. G. 1981. Hostory of botanical science. Academic Press, London.

Müller, I., B. Schmid, and J. Weiner. 2000. The effect of nutrient availability on biomass allocation patterns in 27 species of herbaceous plants. Perspectives in Plant Ecology, Evolution and Systematics **3/2**: 115-127.

Munns, R. 2002. Salinity, growth and phytohormones. Pages 271-290 *in* A. Läuchli and U. Lüttge, editors. Salinity: environment - plants - molecules. Kluwer Academic Publishers, Dordrecht.

Munns, R. and M. Tester. 2008. Mechanisms of salinity tolerance. Annual Review of Plant Biology **59**: 651-681.

Murphy, J. and J. P. Riley. 1962. A modified single solution method for the determination of phosphate in natural waters. Analytica Chimica Acta **27**: 31-36.

Murtagh, F. 1985. Multidimensional Clustering Algorithms. COMPSTAT Lectures 4. Physica-Verlag, Wuerzburg.

Naeem, S. and E. J. Wright. 2003. Disentangling biodiversity effects on ecosystem functioning: Deriving solutions to a seemingly insurmountable problem. Ecology Letters **6**: 567-579.

Niklas, K. J. 1994a. Plant allometry. The scaling of form and process. The University of Chicago Press, Chicago and London.

Niklas, K. J. 1994b. Size-Dependent Variations in Plant Growth Rates and the "3/4 Power Rule". American Journal of Botany **81**: 134-144.

Niklas, K. J. 2006. Plant allometry, leaf nitrogen and phosphorus stoichiometry, and interspecific trends in annual growth rates. Annals of Botany **97**: 155-163.

Niklas, K. J. and E. D. Cobb. 2005. N, P, and C stiochiometry of *Erantis hyemalis* (Ranunculaceae) and the allometry of plant growth. American Journal of Botany **92**: 1256-1263.

Niklas, K. J. and B. J. Enquist. 2001. Invariant scaling relationships for interspecific plant biomass production rates and body size. Proceedings of the National Academy of Sciences of the United States of America **98**: 2922-2927.

Niklas, K. J. and B. J. Enquist. 2002. On the vegetative biomass partitioning of seed plant leaves, stems, and roots. The American Naturalist **159**: 482-497.

Niklas, K. J. and B. J. Enquist. 2003. An allometric model for seed plant reproduction. Evolutionary Ecology Research **5**: 79-88.

Niklas, K. J., T. Owens, P. B. Reich, and E. D. Cobb. 2005. Nitrogen/phosphorus leaf stoichiometry and the scaling of plant growth. Ecology Letters **8**: 636-642.

Obeso, J. R. 2002. The costs of reproduction in plants. New Phytologist **155**: 321-348.

Olff, H., J. de Leeuw, J. P. Bakker, R. J. Platerink, H. J. van Wijnen, and W. de Munck. 1997. Vegetation succession and herbivory in a salt marsh: changes induced by sea level rise and silt deposition along an elevation gradient. Journal of Ecology **85**: 799-814.

Ordoñez, J. C., P. M. van Bodegom, J.-P. M. Witte, I. J. Wright, P. B. Reich, and R. Aerts. 2009. A global study of relationships between leaf traits, climate and soil measures of nutrient fertility. Global Ecology and Biogeography **18**: 137-149.

Osmond, C. B., O. Björkman, and D. J. Anderson. 1980. Physiological Processes in Plant Ecology - Towards a synthesis with *Atriplex*. Springer Verlag, New York.

Packham, J. R. and A. J. Willis. 1997. Ecology of Dunes, Salt Marsh and Shingle. Chapman & Hall.

Paine, R. T. 1969. A note on trophic complexity and community stability. American Naturalist **103**: 91-93.

Pakeman, R. J. 2004. Consistency of plant species and trait responses to grazing along a productivity gradient: a multi-site analysis. Journal of Ecology **92**: 893-905.

Pennings, S. C. and R. M. Callaway. 1992. Salt marsh plant zonation: The relative importance of competition and physical factors. Ecology **73**: 681-690.

Perillo, G. M. E., E. Wolanski, D. R. Cahoon, and M. M. Brinson. 2009. Coastal Wetlands - an integrated ecosystem approach. Elsevier, Amsterdam.

Peters, R. H. 1983. The ecological implications of body size. Cambridge University Press, Cambridge.

Phoenix, G. K., R. E. Booth, J. R. Leake, D. J. Read, J. P. Grime, and J. A. Lee. 2003. Simulated pollutant nitrogen deposition increases P demand and enhances root-surface phosphatase activities of the three functional types in a calcareous grassland. New Phytologist **161**: 279-289.

Pielou, E. C. and R. D. Routledge. 1976. Salt marsh vegetation: Latitudinal gradients in the zonation patterns. Oecologia **24**: 311-321.

Pimm, S. L., G. J. Russell, J. L. Gittleman, and T. M. Brooks. 1995. The future of biodiversity. Science **269**: 347-350.

Pontes, L. D. S., J.-F. Soussana, F. Louault, D. Andueza, and P. Carrere. 2007. Leaf traits affect the above-ground productivity and quality of pasture grasses. Functional Ecology **21**: 844-853.

Poorter, H. and R. de Jong. 1999. A comparison of specific leaf area, chemical composition and leaf construction costs of field plants from 15 habitats differing in productivity. New Phytologist **143**: 163-176.

Poorter, H. and O. Nagel. 2000. The role of biomass allocation in the growth response of plants to different levels of light, CO_2, nutrients and water: a quantitative review. Australian Journal of Botany **27**: 595-607.

Poorter, H. and C. Remkes. 1990. Leaf area ratio and net assimilation rate of 24 wild species differing in relative growth rate. Oecologia **83**: 553-559.

Pott, R. 1995. Farbatlas Nordseeküste und Nordseeinseln. Ulmer Verlag, Stuttgart.

Power, M. E., D. Tilman, J. A. Estes, B. A. Menge, W. J. Bond, L. S. Mills, G. Daily, J. C. Castilla, J. Lubchenco, and R. T. Paine. 1996. Challenges in the quest for keystones. Bioscience **46**: 609-620.

Ranwell, D. S. 1964. *Spartina* salt marshes in southern England: III. Rates of establishment, succession and nutrient supply at Bridgewater Bay, Somerset. Journal of Ecology **52**: 95-105.

Raunkiaer, C. 1907. Planterigets livsformer og deres betydning for geografien Munksgaard, Copenhagen.

Raunkiaer, C. 1934. The life-forms of plants and statistical plant geography. Oxford University Press, Oxford.

Reich, P. B. and J. Oleksyn. 2004. Global patterns of plant leaf N and P in relation to temperature and latitude. Proceedings of the National Academy of Sciences of the United States of America **101**: 11001-11006.

Reich, P. B., C. Uhl, M. B. Walters, and D. S. Ellsworth. 1991. Leaf lifespan as a determinant of leaf structure and function among 23 amazonian tree species. Oecologia **86**: 16-24.

Reich, P. B., M. B. Walters, and D. S. Ellsworth. 1997. From tropics to tundra: Global convergence in plant functioning. Proceedings of the National Academy of Sciences of the United States of America **94**: 13730-13734.

Reineck, H.-E. 1987. Morphologische Entwicklung der Insel Mellum.*in* G. W. Gerdes, E. Krumbein, and H.-E. Reineck, editors. Mellum - Portrait einer Insel, Frankfurt am Main.

Rhodes, D. and A. D. Hanson. 1993. Quaternary ammonium and tertiary sulfonium compounds in higher plants. Annual Review of Plant Physiology and Plant Molecular Biology **44**: 357-384.

Rhodes, D., A. Nadolska-Orczyk, and P. J. Rich. 2002. Salinity, osmolytes, and compatible solutes.*in* A. Läuchli and U. Lüttge, editors. Salinity: environment-plant-molecules. Kluwer, Dordrecht, The Netherlands.

Ribera, I., S. Dolédec, I. S. Downie, and G. N. Foster. 2001. Effect of land disturbance and stress on species traits of ground beetle assemblages. Ecology **82**: 1112-1129.

Richardson, C. J. 1994. Ecological functions and human values in wetlands: a framework for assessing forestry impacts. Wetlands **14**: 1-9.

Root, R. B. 1967. The niche exploitation pattern of the blue-gray gnatcatcher. Ecological Monographs **37**: 317-350.

Rozema, J., P. Bijwaard, G. Prast, and R. Broekman. 1985. Ecophysiological adaptations of coastal halophytes from foredunes and salt marshes. Vegetatio **62**: 499-521.

Rozema, J., H. Gude, F. Bijl, and H. Wesselman. 1981. Sodium concentration in xylem sap in relation to ion exclusion, accumulation and secretion in halophytes. Acta Botanica Neerlandica **30**: 309-311.

Rozema, J., Y. Maanen, H. van Vugts, and A. Leusink. 1983. Airborne and soilborne salinity and the distribution of coastal and inland species of the genus *Elytrigia*. Acta Botanica Neerlandica **32**: 447-456.

Russell, P. J., F. J. Flowers, and M. J. Hutchings. 1985. Comparison of niche breadths and overlaps of halophytes on salt marshes of differing diversity. Vegetatio **61**: 171-178.

Santa Regina, I., M. Rico, M. Rapp, and H. A. Gallego. 1997. Seasonal variation in nutrient concentration in leaves and branches of *Quercus pyrenaica*. Journal of Vegetation Science **8**: 651-654.

Schachtman, D. P., R. J. Reid, and S. M. Ayling. 1998. Phosphorus uptake by plants: from soil to cell. Plant Physiology **116**: 447-453.

Schirmer, U. and S. W. Breckle. 1982. The role of bladders for salt removal in some Chenopodiaceae (mainly *Atriplex* species). Pages 215-231 *in* D. N. Sen and K. S. Rajpurohit, editors. Contributions to the Ecology of Halophytes. Dr Junk, The Hague.

Schlesinger, W. H. 1997. Biogeochemistry: An Analysis of Global Change. Academic, San Diego.

Schlichting, E., H. P. Blume, and K. Stahr. 1995. Bodenkundliches Praktikum. Blackwell, Berlin.

Schmidt-Nielsen, K. 1984. Scaling: why is animal size so important? Cambridge University Press, Cambridge.

Scholten, M., P. A. Blaauw, M. Stroetenga, and J. Rozema. 1987. The impact of competitive interactions on the growth and distribution of plant species in salt marshes. Pages 270-283 *in* A. H. L. Huiskes, C. W. P. M. Blom, and J. Rozema, editors. Vegetation between land and sea Dr. W. Junk, Dordrecht, The Netherlands.

Schulze, E.-D., E. Beck, and K. Müller-Hohenstein. 2002. Pflanzenökologie. Spektrum Akademischer Verlag, Heidelberg, Germany.

Scurlock, J. M. O., K. Johnson, and R. J. Olson. 2002. Estimating net primary productivity from grassland biomass dynamics measurements. Global Change Biology **8**: 736-753.

Sekmen, A. H., I. Türkan, and S. Takio. 2007. Differential responses of antioxidative enzymes and lipid peroxidation to salt stress in salt-tolerant *Plantago maritima* and salt-sensitive *Plantago media*. Physiologia Plantarum **131**: 399-411.

Shipley, B. 2004. Analysing the allometry of multiple interacting traits. Perspectives in Plant Ecology, Evolution and Systematics **6**: 235-241.

Shipley, B. 2010. From Plant Traits to Vegetation Structure: Changes and Selection in the Assembly of Ecological Communities. Cambridge University Press, Cambridge.

Shirley, H. L. 1929. The influence of light intensity and light quality upon the growth of plants. American Journal of Botany **16**: 354-390.

Simberloff, D. 1991. The guild concept and the structure of ecological communities. Annual Review of Ecology and Systematics **22**: 115-143.

Skogley, E. O. and A. Dobermann. 1996. Synthetic ion-exchange resins: soil and environmental studies. Journal of Environmental Quality **25**: 13-24.

Small, C. G. 1996. The Statistical Theory of Shape. Springer, New York.

Smith, T. and M. Huston. 1989. A theory of the spatial and temporal dynamics of plant communities. Vegetatio **83**: 49-69.

Snow, A. A. and S. W. Vince. 1984. Plant zonation in an Alaskan salt marsh. II. An experimental study on the role of edaphic conditions. Journal of Ecology **72**: 669-684.

Sokal, R. and J. Rohlf. 1994. Biometry: the principles and practice of statistics in biological research. W.H. Freeman and Company, New York, USA.

Spalke, J. 2008. Biomasse- und Nährstoffallokation von Dünen- und Salzwiesenpflanzen in Bezug zu ihrer Umwelt - eine Analyse auf der Basis biologischer Pflanzenmerkmale. Diploma thesis. University of Oldenburg, Oldenburg.

Stahl, J. 2002. Foraging along a salinity gradient - the effect of tidal inundation on site choice by dark-bellied brent geese *Branta bernicla* and barnacle geese *B. leucopsis*. Ardea **90**: 201212.

Steadman, D. W. 1995. Prehistoric extinctions of pacific island birds: biodiversity meets zooarcheology. Science **267**: 1123-1131.

Stearns, S. C. 1992. The evolution of Life Histories. Oxford University Press, Oxford.

Stelzer, R. and A. Läuchli. 1977. Salt tolerance and flooding tolerance of *Puccinellia peisonis*. 1. Effect of NaCl-salinity and KCl-salinity on growth at varied oxygen-supply to root. Zeitschrift für Pflanzenphysiologie **83**: 35-42.

Sterner, R. W. and J. J. Elser. 2002. Ecological Stoichiometry: the biology of elements from molecules to the biosphere. Princeton University Press, Princeton, New Jersey.

Steward, G. R., F. Larher, I. Ahmand, and J. A. Lee. 1979. Nitrogen metabolism and salt tolerance in higher plants halophytes. Pages 229-241 *in* R. L. Jefferies and A. J. Davy, editors. Ecological processes in coastal environments. Blackwell, Oxford.

Stoop, J. M. H., J. D. Williamson, and D. M. Pharr. 2006. Mannitol metabolism in plants: a method for coping with stress. Trends in Plant Science **1**: 139-144.

Suding, K. N., S. Lavorel, F. S. Chapin III, J. H. C. Cornelissen, S. Díaz, E. Garnier, D. E. Goldberg, D. U. Hooper, S. T. Jackson, and M. L. Navas. 2008. Scaling environmental change through the community-level: a trait-based response-and-effect framework for plants. Global Change Biology **14**: 1125-1140.

SynBioSys Species Checklist. 2010.
http://www.synbiosys.alterra.nl/synbiosyseu/speciesviewframe.htm Accessed: 07-06-2010.

Tarczynski, M. C., R. G. Jensen, and H. J. Bohnert. 1993. Stress protection of transgenic tobacco by production of the osmolyte mannitol. Science **259**: 508-510.

Teal, J. M. 1986. The ecology of regularly flooded salt marsh of New England: a community profile. U.S. Fish and Wildlife Service. Biol. Report.

Tezara, W., V. J. Mitchell, S. D. Driscoll, and D. W. Lawlor. 1999. Water stress inhibits plant photosynthesis by decreasing coupling factor and ATP. Nature **401**: 914-917.

The R Foundation for Statistical Computing. 2008. R version 2.8.1.

Thioulouse, J., D. Chessel, S. Dolédec, and J. M. Olivier. 1997. ADE4-: a multivariate analysis and graphical display software. Statistics and Computing **7**: 75-83.

Thuiller, W., D. M. Richardson, M. Rouget, S. Procheş, and J. R. U. Wilson. 2006. Interactions between environment, species traits, and human uses describe patterns of plant invasions. Ecology **87**: 1755-1769.

Tilman, D. 1988. Plant Strategies and the Dynamics and Structure of Plant Communities. Princeton University Press, Princeton.

Timmermann, H. 2008. Analyse der Habitatansprüche von Salzwiesen- und Dünenpflanzen auf Mellum. Diploma thesis. University of Oldenburg, Oldenburg.

Tipirdamaz, R., D. Gagneul, C. Duhazé, A. Aïnouche, C. Monnier, D. Özkum, and F. Larher. 2006. Clustering of halophytes from an inland salt marsh in Turkey according to their ability to accumulate sodium and nitrogenous osmolytes. Environmental and Experimental Botany **57**: 139-153.

Tomassen, H. B. M., A. J. P. Smolders, J. Limpens, L. P. M. Lamers, and J. G. M. Roelofs. 2004. Expansion of invasive species on ombrotrophic bogs: desiccation or higher N deposition? Journal of Applied Ecology **41**: 139-150.

Tremp, H. 2005. Aufnahme und Analyse vegetationsökologischer Daten. Ulmer, Stuttgart.

Tyler, G. 1971. Distribution and turnover of organic matter and minerals in a shore meadow ecosystem. Studies in the ecology of baltic sea-shore meadows IV. Oikos **22**: 265-291.

Ungar, I. A. 1991. Ecophysiology of vascular halophytes. CRC Press, Boca Raton, Florida.

Ungar, I. A. 1998. Are biotic factors significant in influencing the distribution of halophytes in saline habitats. Botanical Review **64**: 176-199.

Usuda, H. 1995. Phosphate deficiency in maize. V. Mobilization of nitrogen and phosphorus within shoots of young plants and its relationship to senescence. Plant & Cell Physiology **36**: 1041-1049.

Valladares, F., E. Martinez-Ferri, L. Balaguer, E. Perez-Corona, and E. Manrique. 2000. Low leaf-level response to light and nutrients in Mediterranean evergreen oaks: a conservative resource-use strategy? New Phytologist **148**: 79-91.

van Cleve, K., F. S. Chapin III, C. T. Dyrness, and L. A. Viereck. 1991. Element cycling in taiga forests: state-factor control. Bioscience **41**: 78-88.

van der Maarel, E. and M. van der Maarel-Versluys. 1996. Distribution and conservation status of littoral vascular plant species along the European coast. Journal of Coastal Conservation **2**: 73-92.

Van der Wal, R., M. Egas, A. van der Veen, and J. P. Bakker. 2000. Effects of resource competition and herbivory on plant performance along a natural productivity gradient. Journal of Ecology **88**: 317-330.

van der Werf, A. and O. W. Nagel. 1996. Carbon allocation to shoots and roots in relation to nitrogen supply is mediated by cytokinis and suchrose: opinion. Plant and Soil **185**: 21-32.

Van Diggelen, J. 1991. Effects of inundation stress on salt marsh halophytes. Kluwer Academic Publishers, Dordrecht.

Van Diggelen, J., J. Rozema, D. M. Dickson, and R. Broekman. 1986. β-3-Dimethylsulphoniopropionate, proline and quaternary ammonium compounds in *Spartina anglica* in relation to sodium chloride, nitrogen and sulfur. New Phytologist **103**: 573-586.

van Eerdt, M. M. 1985. The Influence of vegetation on erosion and accretion in salt marshes of the Oosterschelde, the Netherlands. Vegetatio **62**: 327-375.

van Wijnen, H. J. and J. P. Bakker. 1997. Nitrogen accumulation and plant species replacement in three salt-marsh systems in the Wadden Sea. Journal of Coastal Conservation **3**: 19-26.

Vendramini, F., S. Díaz, D. E. Gurvich, P. J. Wilson, K. Thompson, and J. G. Hodgson. 2002. Leaf traits as indicators of resource-use strategy in floras with succulent species. New Phytologist **154**: 147-157.

Vile, D., B. Shipley, and E. Garnier. 2006. A structural equation model to integrate changes in functional strategies during old-field succession. Ecology **87**: 504-517.

Violle, C., M.-L. Navas, D. Vile, E. Kazakou, C. Fortunell, I. Hummel, and E. Garner. 2007. Let the concept of trait be functional! Oikos **116**: 882-892.

Vitousek, P. M. 1982. Nutrient cycling and nutrient use efficiency. American Naturalist **119**: 553-572.

Vitousek, P. M. and R. W. Howarth. 1991. Nitrogen limitation on land and in the sea: How can it occur? Biogeochemistry **13**: 87-115.

Vitousek, P. M., H. A. Mooney, J. Lubchenco, and J. M. Melillo. 1997. Human domination of earth's ecosystems. Science **277**: 494-499.

Vitousek, P. M. and L. R. Walker. 1989. Biological invasion by *Myrica faya* in Hawai'i: plant demography, nitrogen fixation, ecosystem effects. Ecological Monographs **59**: 247-265.

Vrede, T., D. R. Dobberfuhl, S. Kooijman, and J. J. Elser. 2004. Fundamental connections among organism C : N : P stoichiometry, macromolecular composition, and growth. Ecology **85**: 1217-1229.

Walker, B., A. Kinzig, and J. Langridge. 1999. Plant attribute diversity, resilience, and ecosystem function: The nature and significance of dominant and minor species. Ecosystems **2**: 95-113.

Wardle, D. A., G. Hörnberg, O. Zackrisson, M. Kalela-Brundin, and D. A. Coomes. 2003. Long-term effects of wildfire on ecosystem properties across an island area gradient. Science **300**: 972-975.

Warton, D. I., I. J. Wright, D. S. Falster, and M. Westoby. 2006b. Bivariate line-fitting methods for allometry. Biological Reviews **81**: 259-291.

Weiher, E., A. van der Werf, K. Thompson, M. Roderick, E. Garnier, and O. Eriksson. 1999. Challenging Theophrastus: A common core list of plant traits for functional ecology. Journal of Vegetation Science **10**: 609-620.

Weiner, J. 2004. Allocation, plasticity and allometry in plants. Perspectives in Plant Ecology, Evolution and Systematics **6**: 207-215.

Weiner, J., L. G. Campbell, J. Pino, and L. Echarte. 2009. The allometry of reproduction within plant populations. Journal of Ecology **97**: 1220-1233.

Wells, C. G., J. R. Craig, M. B. Kane, and H. L. Allen. 1986. Foliar and soil tests for the prediction of phosphorus response in Loblolly-pine. Soil Science Society of America Journal **50**: 1330-1335.

West, G. B., J. H. Brown, and B. J. Enquist. 1997. A general model for the origin of allometric scaling laws in biology. Science **276**: 122-126.

West, G. B., J. H. Brown, and B. J. Enquist. 1999. The fourth dimension of life: Fractal geometry and allometric scaling of organisms. Science **284**: 1677-1679.

Westoby, M. 1998. A leaf-hight-seed (LHS) plant ecology strategy scheme. Plant and Soil **199**: 213-227.

Westoby, M., D. S. Falster, A. T. Moles, P. A. Vesk, and I. J. Wright. 2002. Plant ecological strategies: Some leading dimensions of variation between species. Annual Review of Ecology and Systematics **33**: 125-159.

Wiebe, H. H. 1978. The significance of plant vacuoles. Bioscience **28**: 327-331.

Wiehe, P. O. 1935. A quatitative study of the influence of tide upon populations of *Salicornia euopea* Journal of Ecology **23**: 323-333.

Williams, M. D. and I. A. Ungar. 1972. The effect of environmental parameters on the germination, growth, and development of *Suaeda depressa* (Pursh) Wats. The American Naturalist **59**: 912-918.

Wilson, J. B. 1999. Guilds, functional types and ecological groups. Oikos **86**: 507-522.

Wolters, M., A. Garbutt, R. M. Bekker, J. P. Bakker, and P. D. Carey. 2008. Restoration of salt-marsh vegetation in relation to site suitability, species pool and dispersal traits. Journal of Applied Ecology **45**: 908-912.

Wright, E. J. and K. Cannon. 2001. Relationships between leaf lifespan and structural defences in a low-nutrient, sclerophyll flora. Functional Ecology **15**: 351-359.

Wright, I. J., P. B. Reich, and M. Westoby. 2001. Strategy shifts in leaf physiology, structure and nutrient content between species of high- and low-rainfall and high- and low-nutrient habitats. Functional Ecology **15**: 423-434.

Wright, I. J., P. B. Reich, M. Westoby, D. D. Ackerly, Z. Baruch, F. Bongers, J. Cavenders-Bares, T. Chapin, J. H. C. Cornelissen, M. Diemer, J. Flexas, E. Garnier, P. K. Groom, J. Gulias, K. Hikosaka, B. B. Lamont, T. Lee, W. Lee, C. Lusk, J. J. Midgley, M.-L. Navas, Ü. Niinements, J. Oleksyn, N. Osada, H. Poorter, P. Poot, L. Prior, V. I. Pyankow, C. Roumet, S. C. Thomas, M. G. Tjoelker, E. J. Veneklaas, and R. Villar. 2004. The worldwide leaf economics spectrum. Nature **428**: 821-827.

Wyn Jones, G. and J. Gorham. 2002. Intra- and inter-cellular compartments of ions. Pages 159-180 *in* A. Läuchli and U. Lüttge, editors. Salinity: environment-plant-molecules. Kluwer, Dordrecht, The Netherlands.

Yachi, S. and M. Loreau. 1999. Biodiversity and ecosystem productivity in a fluctuating environment: The insurance hypothesis. Proceedings of the National Academy of Sciences of the United States of America **96**: 1463-1468.

Zimov, S. A., V. I. Chuprynin, A. P. Oreshko, F. S. I. Chapin, J. F. Reynolds, and M. C. Chapin. 1995. Steppe-tundra transition: a herbivore-driven biome shift at the end of the pleistocene. The American Naturalist **146**: 765-794.

Appendix

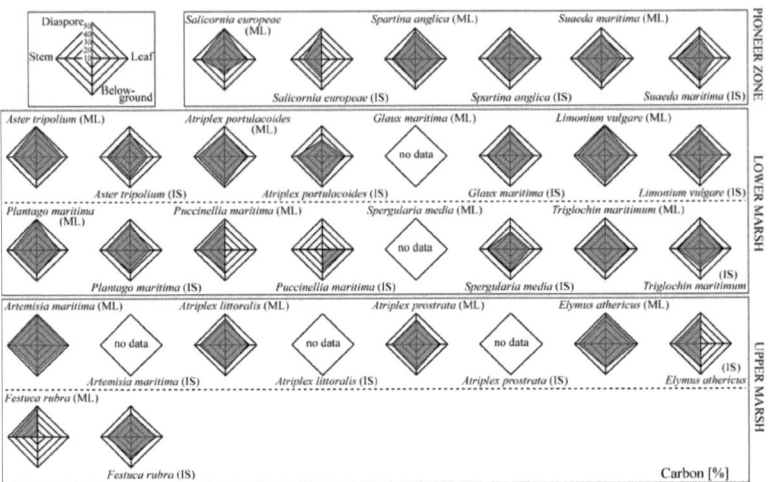

Appendix 1: Radar chart of carbon content [%] in organs of salt marsh plants (diaspore, stem, leaf and belowground). Plant species were arranged to refer to pioneer zone, lower and upper marsh and separated with respect to appearance on mainland (ML) or island (IS) marshes.

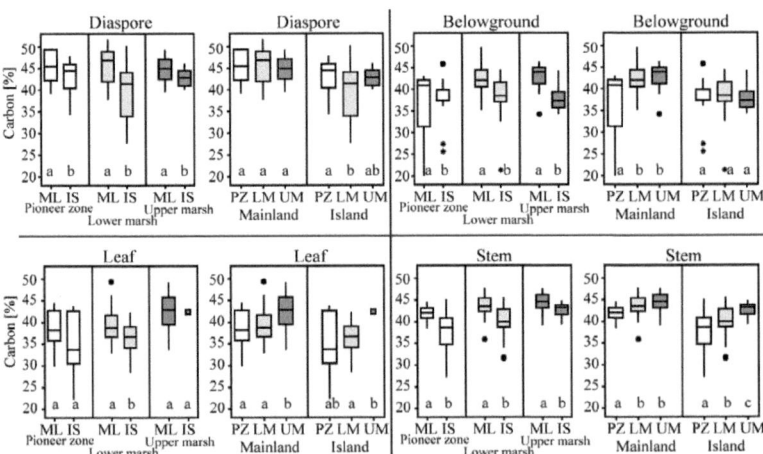

Appendix 1: Boxplots of carbon content [%] of plant organs (diaspores, belowground, leaf and stem) for mainland and island species of the pioneer zone, lower and upper marsh on mainland and island site, respectively. Homogenous subgroups as results of t- and H-Test are shown by the use of the same letters (p-value > 0.05). Circles indicate outside values (sample point > upper quartile + 1.5 times distance to quartile), asterisks far outside values (sample point > upper quartile + 3 times distance to quartile).

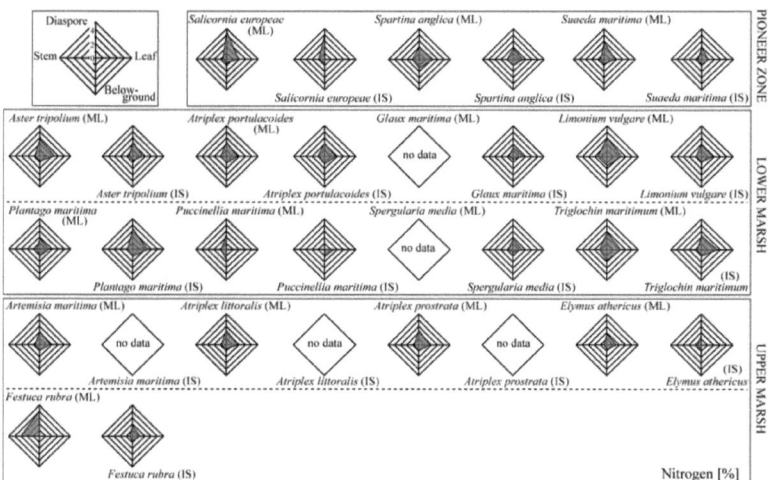

Appendix 2: Radar chart of nitrogen content [%] in organs of salt marsh plants (diaspore, stem, leaf and belowground). Plant species were arranged to refer to pioneer zone, lower and upper marsh and separated with respect to appearance on mainland (ML) or island (IS) marshes.

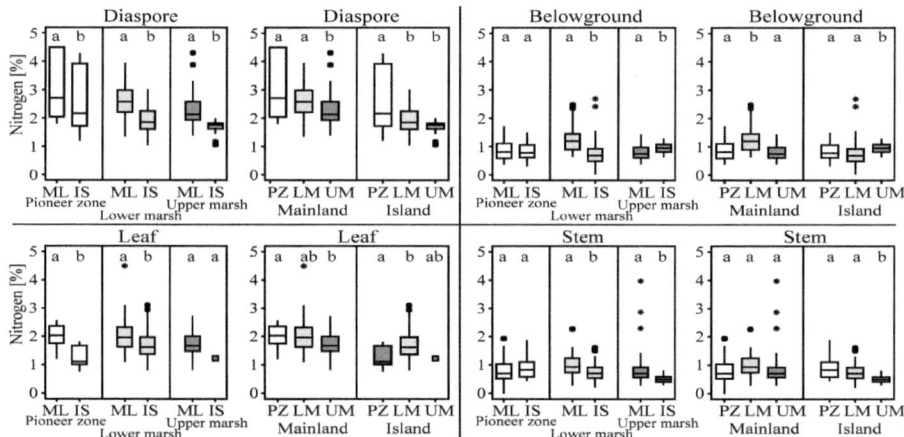

Appendix 3: Boxplots of nitrogen content [%] of plant organs (diaspores, belowground, leaf and stem) for mainland and island species of the pioneer zone, lower and upper marsh on mainland and island site, respectively. Homogenous subgroups as results of t- and H-Test are shown by the use of the same letters (p-value > 0.05). Circles indicate outside values (sample point > upper quartile + 1.5 times distance to quartile), asterisks far outside values (sample point > upper quartile + 3 times distance to quartile).

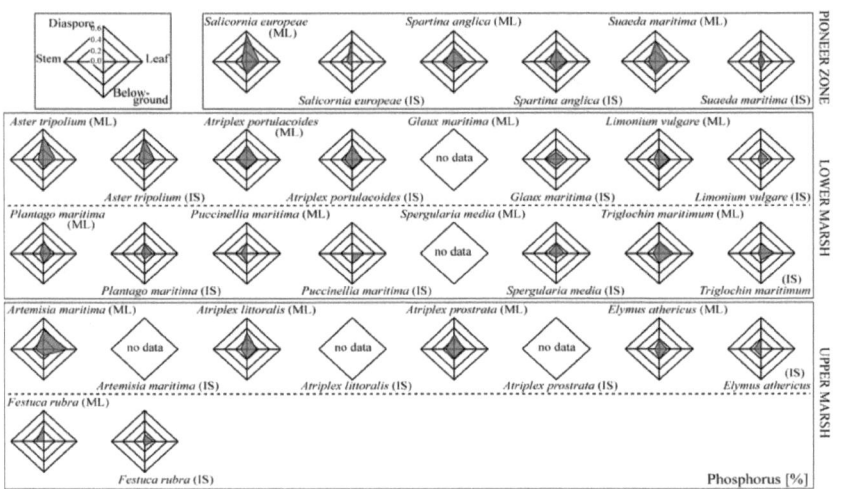

Appendix 4: Radar chart of phosphorus content [%] in organs of salt marsh plants (diaspore, stem, leaf and belowground). Plant species were arranged to refer to pioneer zone, lower and upper marsh and separated with respect to appearance on mainland or mainland (ML) or island (IS) marshes.

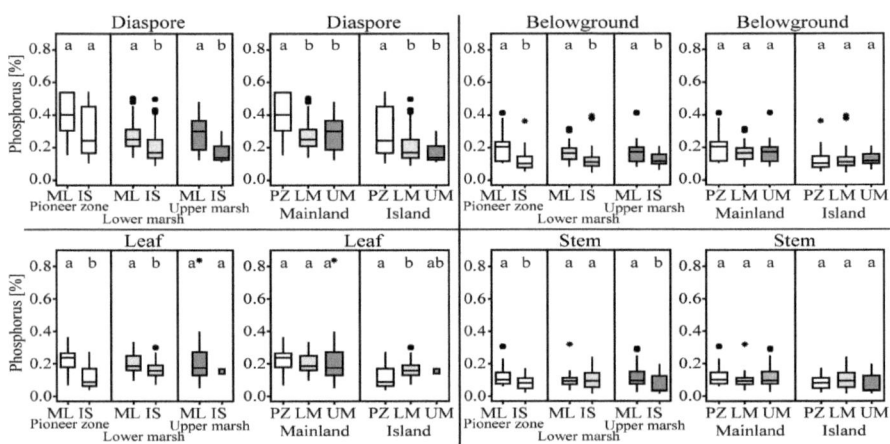

Appendix 5: Boxplots of phosphorus content [%] of plant organs (diaspores, belowground, leaf and stem) for mainland and island species of the pioneer zone, lower and upper marsh on mainland and island site, respectively. Homogenous subgroups as results of t- and H-Test are shown by the use of the same letters (p-value > 0.05). Circles indicate outside values (sample point > upper quartile + 1.5 times distance to quartile), asterisks far outside values (sample point > upper quartile + 3 times distance to quartile).

Appendix 6: Significant correlations as result of fourth-corner analysis for data obtained on mainland marshes. All relationships are at p≤ 0.05 and are either positive or negative. Number in cells are correlation coefficients (r), empty cells indicate no significant correlation. Correlation coefficients > ±0.20 are shown in bold. For abbreviations see ‚List of most important abbreviations' and Table 5-1. Phosp.=Phosphorus, Pot.=Potassium, Carb.=Carbonate, Inun.=Inundation frequency, GW.mean = mean level of groundwater, GW.sal = salinity content of groundwater.

MAINLAND	Phosp.	Pot.	Carb.	Sand	Inun.	GW.mean	GW.sal
Dilution					0.19	0.13	0.09
Exclusion			-0.22		0.16	**0.37**	**0.28**
MAOS			**0.20**		**-0.25**	**-0.43**	**-0.35**
Canopy height				-0.07	0.10	0.14	
SLA			-0.12		0.11	**0.22**	0.08
LDMC			0.17		**-0.25**	**-0.36**	**-0.27**
SDMC			0.08		-0.18	**-0.22**	-0.08
RE					0.08		-0.09
SMF					-0.10		**-0.21**
LMF			0.10			-0.18	
RMF			-0.15		0.15	**0.24**	**0.33**
Dry weight of							
stem						0.07	
leaf						0.06	
diaspores							
belowground			-0.12			0.19	0.16
C:N ratio of							
whole plant			0.09				
stem	0.11		**0.22**		**-0.22**	**-0.35**	**-0.26**
leaf	0.08		0.19		**-0.28**	**-0.38**	**-0.30**
diaspores	0.10		0.19		**-0.21**	**-0.28**	-0.19
belowground		-0.09	0.18		-0.17	**-0.30**	**-0.31**
C:P ratio of							
stem	0.10	-0.08	**0.20**		**-0.20**	**-0.35**	**-0.25**
leaf	0.08		0.19		**-0.27**	**-0.39**	**-0.30**
diaspores			0.09		-0.17	-0.19	-0.09
belowground		-0.09	0.13		-0.13	**-0.25**	**-0.26**
N:P ratio of							
stem							
leaf			0.09		-0.19	**-0.28**	**-0.21**
diaspores				0.08			
belowground			-0.15	0.08	0.14	0.07	0.13

Appendix 7: Significant correlations as result of fourth-corner analysis for data obtained on the island of Mellum. All relationships are at p≤ 0.05 and are either positive or negative. Number in cells are correlation coefficients (r), empty cells indicate no significant correlation. Correlation coefficients > ±0.20 are shown in bold. For abbreviations see ‚List of most important abbreviations' and Table 5-1. Phosp.=Phosphorus, Pot.=Potassium, Carb.=Carbonate, Inun.=Inundation frequency, GW.mean = mean level of groundwater, GW.sal = salinity content of groundwater.

ISLAND	Phosp.	Pot.	Carb.	Sand	Inun.	GW.mean	GW.sal
Dilution	0.17					**0.27**	**0.22**
Exclusion	**0.21**	0.18	**-0.20**			**0.22**	**0.35**
MAOS	**0.42**	**0.31**	**0.30**			**-0.53**	**-0.64**
Canopy height						-0.19	
SLA							-0.17
LDMC	**-0.37**	**-0.33**	**0.27**			**-0.43**	**-0.57**
SDMC	**-0.37**	-0.19	**0.28**			-0.18	**-0.49**
RE		-0.16	0.17				**-0.34**
SMF		**0.21**				**-0.52**	
LMF	**0.29**	**0.21**		-0.16		**0.39**	**0.44**
RMF							
Dry weight of							
stem		**0.24**		-0.15			**0.22**
leaf			-0.14				0.19
diaspores			-0.14				0.17
belowground			-0.14				0.13
C:N ratio of							
whole plant	0.18					0.17	0.15
stem	**-0.31**	**-0.26**	**0.29**	-0.15		**-0.42**	**-0.54**
leaf							**-0.25**
diaspores	**-0.33**	**-0.35**	**0.31**			**-0.34**	**-0.55**
belowground			0.12				
C:P ratio of							
stem			**0.22**	-0.14		-0.19	**-0.31**
leaf						0.14	
diaspores	-0.17	**-0.27**	**0.25**				**-0.31**
belowground						0.19	0.14
N:P ratio of							
stem							
leaf						0.18	**0.25**
diaspores							
belowground	0.18	0.17	-0.13			**0.27**	**0.34**

Appendix 8: Significant correlations as result of fourth-corner analysis of the whole dataset. All relationships are at p≤ 0.05 and are either positive or negative. Number in cells are correlation coefficients (r), empty cells indicate no significant correlation. Correlation coefficients > ±0.20 are shown in bold. For abbreviations see 'List of most important abbreviations' and Table 5-1. Phosp.=Phosphorus, Pot.=Potassium, Carb.=Carbonate, Inun.=Inundation frequency, GW.mean = mean level of groundwater, GW.sal = salinity content of groundwater.

ALL DATA	Phosp.	Pot.	Carb.	Sand	Inun.	GW.mean	GW.sal
Dilution		-0.11	-0.13		0.13	**0.21**	0.14
Exclusion		0.15		-0.10	0.13	**0.28**	**0.27**
MAOS			0.16		**-0.22**	**-0.45**	**-0.41**
Canopy height	**0.26**	**0.34**	**0.39**	**-0.29**		0.07	-0.09
SLA		0.10		-0.07		0.08	
LDMC	0.15	0.12	**0.31**	-0.10	**-0.23**	**-0.39**	**-0.32**
SDMC	-0.10	-0.08		0.14	-0.15	**-0.26**	-0.14
RE	**0.22**	**0.30**	**0.36**	**-0.20**		-0.16	**-0.20**
SMF	**0.29**	**0.39**	**0.44**	**-0.32**	-0.13	-0.19	-0.16
LMF	-0.18	**-0.31**	**-0.31**	**0.20**		0.08	0.08
RMF	**-0.23**	**-0.29**	**-0.40**	**0.25**	0.14	**0.26**	**0.24**
Dry weight of							
stem	0.13	0.14	0.15	-0.14			
leaf	-0.10	-0.11	-0.18	0.09		0.12	0.10
diaspores	0.11	0.09	0.12	-0.11			
belowground	-0.15	-0.19	**-0.28**	0.16		0.18	0.14
C:N ratio of							
whole plant	-0.13	**-0.22**	**-0.28**	0.14		0.17	0.12
stem	0.14		**0.28**		**-0.22**	**-0.38**	**-0.32**
leaf			0.09		**-0.21**	**-0.29**	**-0.28**
diaspores	-0.08	-0.16		0.11	-0.15	**-0.25**	**-0.25**
belowground		-0.10			-0.14	-0.19	**-0.25**
C:P ratio of							
stem	-0.19	**-0.28**	**-0.21**	**0.23**	-0.11	-0.13	-0.17
leaf			0.07		**-0.21**	**-0.23**	**-0.21**
diaspores		-0.10		0.13	-0.12	-0.14	-0.12
belowground	-0.08	**-0.20**	-0.12	0.11	-0.08		-0.13
N:P ratio of							
stem	**-0.27**	**-0.30**	**-0.36**	**0.26**		0.11	
leaf		-0.07	-0.07		-0.10		
diaspores					0.08		
belowground	**-0.22**	**-0.28**	**-0.42**	**0.23**	0.13	**0.25**	**0.22**

Acknowledgments

There is a whole bunch of people who contributed to this work in one way or another, who scientifically advised and supported me and, maybe even more important, stayed and became friends to me during my PhD-time and reminded me that work is not everything.

My special thanks go to:

Michael Kleyer for being my supervisor and ensuring me that I can write good manuscripts even if the reviewers disagree. I am also grateful to Helmut Hillebrand for accepting to be referee of this thesis.

For financial support, the Wasserverbandstag e.V. and the II. Oldenburgische Deichband. Also the administration of the National Park 'Niedersächsisches Wattenmeer' for their support during field work.

The Landscape Ecology Group of the University of Oldenburg, especially Regine Kayser for her help in the lab and Cord Peppler-Lisbach for his thorough explanations on everything. Brigitte and Helga for providing the right working environment. Martin for being the other half of the TREIBSEL-project and for knowing that salt marshes are a 'special' place to work in.

Special thanks to Jürgen Böhmer for scientific and personal advise and that I will always be the 'young' scientist next to him.

Andrea, Martin and Jürgen for proof reading and comments on the thesis.

Many students who helped me collecting the data, especially Sandra Andratschke, Nora Leipner, Karo Pawletko and Timm Peyrat. Thanks to Sandra Andratschke, Hanna Timmermann and Janina Spalke for allowing me to use their data.

And a very special thanks to all the people who gave me a good time in Oldenburg, who supported me during this time, during the time before and (hopefully) the time to come. You all mean a lot to me and you will know that I speak of you when you read this.

I want morebooks!

Buy your books fast and straightforward online - at one of world's fastest growing online book stores! Environmentally sound due to Print-on-Demand technologies.

Buy your books online at
www.morebooks.shop

Kaufen Sie Ihre Bücher schnell und unkompliziert online – auf einer der am schnellsten wachsenden Buchhandelsplattformen weltweit! Dank Print-On-Demand umwelt- und ressourcenschonend produziert.

Bücher schneller online kaufen
www.morebooks.shop

KS OmniScriptum Publishing
Brivibas gatve 197
LV-1039 Riga, Latvia
Telefax: +371 686 204 55

info@omniscriptum.com
www.omniscriptum.com

Printed by Books on Demand GmbH, Norderstedt / Germany